P9-DMH-638

The Future of Evolution in the Aftermath of Man

The NEXT SPECIES

MICHAEL TENNESEN

SIMON & SCHUSTER
New York London Toronto Sydney New Delhi

Simon & Schuster
1230 Avenue of the Americas
New York, NY 10020

First Simon & Schuster hardcover edition March 2015

Book design by Ellen R. Sasahara

Manufactured in the United States of America

1 3 5 7 9 10 8 6 4 2

Library of Congress Cataloging-in-Publication Data

Tennesen, Michael.
The next species: the future of evolution in the aftermath of man / Michael Tennesen. — First Simon & Schuster hardcover edition.
pages cm
Includes bibliographical references and index.
1. Evolution (Biology) 2. Mass extinctions. 3. Nature—Effect of human beings on. I. Title.
QH366.2.T465 2015
576.8—dc23

2014037267

ISBN 978-1-4516-7751-5
ISBN 978-1-4516-7753-9 (ebook)

To Annabelle, my mother,
who loved the oceans, the mountains, the deserts,
the birds, the animals, and *the people.*

Every part of this soil is sacred in the estimation of my people. Every hillside, every valley, every plain and grove, has been hallowed by some sad or happy event in days long vanished.

—Chief Seattle, 1854

We live in a zoologically impoverished world, from which all the hugest and fiercest and strangest forms have recently disappeared.

—Alfred Russel Wallace, 1876

Contents

Part IV NOW WHAT?

THE GEOLOGIC TIME SCALE

Note: "mya" means "million years ago"

EON	ERA	PERIOD	EPOCH
Phanerozoic (541.0 mya to present)	**Cenozoic** (66.0 mya to present)	**Quaternary** (2.58 mya to present)	**Holocene** (11,700 yrs to present)
			Pleistocene (2.58 mya to 11,700 yrs)
		Neogene (23.03 to 2.58 mya)	
		Paleogene (66.0 to 23.03 mya)	
	Mesozoic (252.0 to 66.0 mya)	**Cretaceous** (145.0 to 66.0 mya)	
		Jurassic (201.3 to 145.0 mya)	
		Triassic (252.0 to 201.3 mya)	
	Paleozoic (541.0 to 252.17 mya)	**Permian** (298.9 to 252.17 mya)	
		Carboniferous (358.9 to 298.9 mya) (**Pennsylvanian** and **Mississippian**)	
		Devonian (419.2 to 358.9 mya)	
		Silurian (443.4 to 419.2 mya)	
		Ordovician (485.4 to 443.4 mya)	
		Cambrian (541.0 to 485.4 mya)	

SUPEREON	EON	PERIOD
Precambrian (4600 to 541.0 mya)	**Proterozoic** (2500 to 541.0 mya)	**Ediacaran** (635.0 to 541.0 mya)
	Archean (4000 to 2500 mya)	
	Hadean (4600 to 4000 mya)	

Prologue

WE HAVE NO IDEA WHAT WE'RE IN FOR

IT WAS MID-MORNING, June, during the tropical dry season, as the Peruvian army Mi-17 helicopter lifted us off from a military base near the town of Ayacucho, Peru, on the western side of the Andes Mountains and slowly ascended toward the crest of the magnificent range. The expansive dry terrain below was spotted with cactus, shrubs, and wide stretches of open space, interrupted only by small villages covered in a fine layer of the local dust.

These slopes constitute the eastern boundary of the Atacama Desert, one of the driest spots on earth. It gave no hint of the verdant rain forest that awaited us just beyond the summit of the Andes. But as the helicopter crested the mountains, the eyes of the passengers—a military crew and an international team of scientists—opened wide at the sudden appearance of the headwaters of the Amazon River and the thick blanket of deep green vegetation that cloaked the mountains on this much wetter terrain.

Inside the helicopter, the group of celebrated biologists, part of the Rapid Assessment Program, had been sent here by the Washington, DC–based environmental group Conservation International to do a quick and dirty survey of the wildlife in the tropical forest region of the

Vilcabamba, one of several mountain ranges within the eastern Andes under threat by oil and mining interests. Conservation International wanted to know if the area was rich enough in the number of plant and animal species to warrant the use of the group's limited funds to save it. The more species there were, the more likely that some would survive the current environmental crisis.

I sat with the scientists on uncomfortable metal benches bolted to the wall, gear piled high around us. Most were dressed in khakis with an assortment of high-top boots, a few beards, and several parkas. They all tried to peer out the cloudy glass portals and the open door of the cabin, excited by their first look at the tropical forest they'd come to study. A Peruvian soldier, wearing no seat belt, one arm hooked through a wall handle adjacent to the open door, was perched dangerously with his legs and gun dangling out the helicopter. Insurgents had wounded one of his comrades the day before, and he scanned the forest below, looking for trouble.

Our view stretched eastward over the Amazon Basin where the sun had already begun to heat the tropical forest, turning its moisture into towering thunderheads, which by noon would begin to assault the eastern face of the Andes with wave upon wave of mist and rain. The result of all this water was a lush tropical menagerie, an area scientists consider to be the most biologically diverse of all the remaining forests on earth. The enormous number of species of fauna and flora in the Andes and in the adjoining Amazon Basin is as vital to the health of the tropics as it is to the world. This area gave birth to many of the terrestrial plants and animals on earth and is thus responsible for much of the world's species diversity—its "biodiversity." Scientists tell us that nature is currently heading toward one of the major catastrophes of its existence, a deadly crisis brought on by the land use activities of man, resulting in the plummet of species numbers. Our best hope and why so many scientists were aboard this helicopter was that the tropics could serve as a repository from which nature could resurrect replacements in the future.

There is reason and precedent for this hope, which is why these scientists are studying this specific landscape: during past ice ages, for example, most Andean animals and plants moved down from the precipices and held out in isolated pockets of rain forest at lower elevations. While glaciers scoured much of the earth, closer to the poles, destroying all life that could not get out of the way, the Andes and the Amazon functioned as a warm safe haven from this frozen assault.

Today, the eastern Andes Mountains is one of the few places on earth where new species, animals not yet discovered by science, still abound. The area is classified as a global hot spot, a terrain with dense biodiversity, featuring many species found nowhere else in the world. It is in areas like this, in dark and difficult corners of the globe that scientists hope nature might survive man's current assault, and new species could reappear.

The mountainous terrain below our helicopter featured an area known as "cloud forest" where trees were shrouded in mosses and ferns. The canopies were filled with orchids and bromeliads that cast their roots into the leaves and humus in the crooks of the trees or into the bark of the branches in place of dirt.

Many of the species here had what Wake Forest University biologist Miles Silman described as "shoestring distributions." The area where they can grow and reproduce may stretch horizontally for hundreds of miles but vertically only a few hundred feet. "I can throw a rock over the elevational distribution of some of these plants," said Silman. He fears that climate change could push species uphill too fast for them to adapt.

There is a reason they call this "cloud forest." It could take several days to land in such an area because of the constant cloud cover. The first day we tried, our army helicopter was turned back by the weather, and the pilot decided to visit the Asháninka Indians instead. Tribal members all came out to greet us. Their faces and arms were streaked with berry juice, a jungle version of makeup. A woman offered us *chicha*, a liquid made from yucca that is masticated and fermented by the tribal women, which the pilot told us to accept, to avoid gravely insulting the community. The Asháninka still took game from the local forests and fish from nearby streams.

On the third day, the clouds finally broke and we landed. I was one of the first people out of the helicopter, and my boots sank deeply into the boggy soil. I turned to the scientist behind me and told her I thought this was the wrong place. But she would have none of my hesitancy. "This is it," she said, and gestured for me to get going. Within hours we'd unloaded the gear and hacked our way with machetes through the forest to a knoll where we cleared an area and set up a functional though very damp camp.

The tropical Andes contain about a sixth of the world's plant life in less than 1 percent of its land area. White-faced monkeys, spider monkeys, and mantled howler monkeys swing through the trees and fill the damp air with their screams and roars. Puma, bear, white-lipped peccaries, and mountain tapirs patrol the woods looking for dinner, while the birds, bats, and butterflies shadow their movements. There are more than 1,724 species of birds in an area the size of New Hampshire—better than double the number found in Canada and the US combined.

The Vilcabamba Range is cut off from the surrounding mountains by the deep valleys of the Apurímac and Urubamba Rivers. Rising like an island in a sea of jungle, it is as isolated as an island surrounded by ocean.

Life is unique in the tropics. Animals often specialize, living off a single plant or groups of plants. Some flowers have long, curved tubes that can only be pollinated by certain species of hummingbirds with similarly curved bills. But there are also cheaters, like the flowerpiercer, a bird that can use its hooked bill like a beer can opener, notching little holes at the bases of flowers so that it, as well as bees and small hummingbirds, can get at the nectar without having to go through the flowers' long tubes.

One night about a week into our trip when the rains started coming down, the resident herpetologist Lily O. Rodríguez and I put on headlamps and headed out into the deluge looking for new species, since rains brought out the different frogs and amphibians. Rodríguez started telling me stories about how these animals learn to specialize in the face of intense competition. She said that some of the frogs here don't have tadpoles; they sit on their eggs like chickens. Other frogs store their tadpoles in leaves above streams into which the tadpoles fall once they hatch. And then some tadpoles have huge mouths to hold on to their favorite rocks when the streams run too fast.

The rains grew heavier and we put on waterproof army ponchos over Gore-Tex parkas. But this didn't stop Rodríguez from climbing out on a wet, slippery tree limb when she thought she heard a new frog croak. She found nothing out on the tree limb that night, but she came across twelve new species in the course of our four-week visit.

The wonder of evolution is exemplified in these rarefied, verdant corners, where life adapts to tiny ecological niches of nature that require elaborate maneuvers for others to take advantage of. The question is: Will nature provide the necessary niches and maneuvers to meet the future? Will the tropics be part of the rescue, if there is one? And will modern man be along for the ride?

The palpable haste of modern biologists is due partly to the fear that we may be at the start of a mass extinction event—a loss of over 75 percent of plant and animal species. Such events have occurred only five times in the past 600 million years, when animals first appeared in the fossil record. And now scientists suggest that a sixth mass extinction may be under way, given the known species losses over the past few centuries and millennia. A recent report in the science journal *Nature* from biologists at the University of California, Berkeley, states that we could reach the extreme of a mass extinction in as little as three centuries from now if current threats to species are not alleviated.

It took *Homo sapiens* less than 200,000 years to reach a burgeoning population of one billion in 1800, but by 2000 we topped off at six billion, and by 2045 we are projected to reach nine billion. It is an unprecedented surge of growth, with unimaginable risks and innumerable side effects—the wellspring of a raging crisis.

And yet it is a dilemma man appears to ignore, though it is becoming more difficult to disregard as the list of earth's endangered plants and animals keeps growing due to our multiple assaults on the environment. We have become a deadly virus to nature.

Our massive overpopulation and accompanying decimation of earth's natural resources, if pursued unabated, may lead to man's own demise. Yet, as great as our footprint has been, from a geological perspective, we've done all this damage in a brief moment. If one looks at the entire history of earth as a twenty-four-hour day, we only entered the picture around the last seconds of that day. We work fast.

Of course, earth will recover, no matter how devastating our brief visit here. After all, just because it may mean the end of man, it won't be the demise of all biological life. Life is resilient. Plants, animals, and microbes will survive, adapt, diversify, and proliferate. New plants will evolve to vanquish our monocultures of corn, wheat, and rice. With far fewer animals around, those species that survive the bottleneck of extinction will move into newly abandoned spaces. With little competition, they will thrive and rapidly evolve.

It's all happened before.

Recoveries followed all the mass extinctions, no matter their causes. The Ordovician extinction event 443 million years ago destroyed 86 percent of species with a barrage of alternating glacial cycles. The Devonian event 359 million years ago took out 75 percent of species with a one-two punch of global cooling and global warming. The Permian event 252 million years ago destroyed 96 percent of species with a Siberian supervolcano. The Triassic event 200 million years ago took out 80 percent of species with a combination of global warming and ocean acidification. The Cretaceous event 65 million years ago destroyed 76 percent of species with the impact of an asteroid. Though we have identified the prime suspects here, each of these mass extinctions had multiple causes.

The best known, the Cretaceous extinction 65 million years ago, was the primary result of an asteroid impact, though it had help from a supervolcano, the Deccan Traps in India—traps being large regions of volcanic rock with step-like plateaus and mountains that are typical of flood basalt eruptions. The Permian extinction 252 million years ago, the child of a volcano, also had help from the collapse of ocean currents, among other causes. Yet, despite their enormous destruction, the Permian extinction opened the door for the dinosaurs, and the Cretaceous extinction opened the door for mammals and man.

Extinction is a powerful creative force, says Douglas H. Erwin, a paleobiologist at the Smithsonian Institution. In his book *Extinction: How Life on Earth Nearly Ended 250 Million Years Ago* he writes, "From the wreckage of mass extinctions the survivors are free for bursts of evolutionary creativity, changing the dominant members of the ecological communities, and enabling life to move off in new and unexpected directions."

Anthony Barnosky, a professor of integrative biology at the University of California, Berkeley, and principal author of the *Nature* paper, says that the critical component in determining if we are headed toward a mass extinction event is the status of critically endangered, endangered, and vulnerable species. "With them, Earth's biodiversity remains in pretty good shape compared to the long-term biodiversity baseline. But if most of them die, even if their disappearance is stretched out over the next 1,000 years, the sixth mass extinction will have arrived," he says.

He thinks that if we save the species now considered in trouble, we may have a chance. But our work at saving endangered species has resulted in many cases of what paleontologists refer to as "dead clade walking"—"clade" meaning groups of organisms. An example of lingering species is the California condor, which is threatened by lead poisoning, lethal pesticides, and expanding urban areas. It has cost millions of dollars and countless hours of work to preserve critical habitat, raise captive birds, and release them to the wild—but will the California condor be here for the next thousand years?

And if so, will other birds survive the bottleneck of mass extinction? Will reptiles, fish, insects, mammals, and perhaps even man survive? And how will they differ from the current versions of their species? That is what we will investigate here.

This book looks at past extinction events, the evolution of man and nature, evolutionary changes already under way, and the evolutionary changes likely to occur. Our title, *The Next Species*, is used in its plural sense. We are interested in the next species of marine and terrestrial animals as much as we are in the next species of man. The research was built on scientific papers, books, as well as personal visits and interviews with experts. Its vision is based on fossil evidence from the past, studies of the present, and expert predictions of the future.

I visited more than seventy scientists from Harvard, MIT, Duke, the Smithsonian Institution, the American Museum of Natural History, UC Berkeley, Stanford, the University of Indiana, the University of London, Oxford University, the Max Planck Institutes, and more. We spoke on the phone with many others.

Many, like Hans-Dieter Sues, curator of vertebrate paleontology at the Smithsonian, think that extinction is a normal process of life. "Virtually 99.999 percent of all life on the planet has gone extinct," says Sues. "And so will *Homo sapiens*. Maybe in one thousand years we will have figured out how to do interstellar travel, so if things go haywire down here, we can take off and go somewhere else. But it's just as possible we will mess around with our own genome and create some sort of race of superhumans, and they'll drive us to extinction."

This book looks around the world for lessons in evolution. What can past mass extinctions teach us? Can pristine ecosystems exist in war zones and nuclear accidents? What can 30,000-year-old fossils under Los Angeles tell us about the diversity of life? Will scientists rewild the Americas and Europe with elephants, cheetahs, and lions? Will jellyfish and giant squid dominate the oceans? Does disease fester in a world devoid of its native species? And what about the chances of an escape to Mars?

We'll also explore the possibility of other forms of life evolving. Could isolation in the wake of a mass extinction provide the evolutionary opportunity for another species of man? Will genetics provide our children with better minds, longer lives, and unique bodies? Or will scientists figure out how to upload the human mind, making our bodies obsolete, so that we live on as robots or avatars in a virtual world.

The possibilities abound.

Part I

VISIT TO THE PAST

1

A MASS EXTINCTION: THE CRIME SCENE

IF YOU'RE CURIOUS as to what a mass extinction looks like, you might want to visit the remains of the Capitan Reef at Guadalupe Mountains National Park, the highest mountains in Texas. Life abounded in the seas back then, but the dinosaurs had yet to appear. The creatures that walked the dry land were not as enormous, nor as diversified, as they would become later. The continents were bound together in a single landmass, but as it broke up and drifted apart, the movement provided the isolation necessary for new species to evolve. Still, life had to sidestep the Permian extinction before it could truly flourish. The story of life's decimation at this point, followed by its resurrection, has multiple lessons for our own predicament.

The Capitan Reef, though long dead, once thrived between 272 and 260 million years ago in the middle of the Permian period, just before the greatest mass extinction the world has ever known. The International Union of Geological Sciences has selected three points within the park as "golden spikes," the standard against which all other rocks of the Middle Permian period are compared. (The actual markers that indicate these points are brass plaques.)

At the bottom of the trail up to the reef one day, I met Guadalupe Mountains National Park geologist Jonena Hearst. After she patiently answered questions from a park visitor and showed me some maps and geological charts, a process that took twenty minutes, she loaded up her day pack and lifted it onto her back. With a big broad smile she signaled it was time to get going. I followed after her. It was fall, a transitional time in West Texas weather, when mornings bore the chill of impending winter, yet the afternoons carried a remembrance of summer heat. McKittrick Canyon before us cut a slash through the Guadalupe Range, exposing the backbone of the Capitan Reef—one of the most extensive fossil reef formations on earth.

The surrounding terrain was dry, open desert, with cactus and creosote brush sheltering an assortment of rabbits, snakes, and lizards—a marked contrast from the tropical rain forests of Vilcabamba. Whereas the rain forest is full of moisture and life, the desert is bashful about any display of exuberance. Farther up McKittrick Canyon, cottonwoods surrounded a portion of the stream that surfaced intermittently. The trees were full of inviting autumn colors, but our path quickly pulled away from the stream and veered up a steep embankment toward the Capitan Reef above us.

What we saw of the reef exposed here displayed the calcified remains of an enormous formation of shelled creatures and sponges that once lay beneath an ancient sea, not unlike the coral reefs of today. A huge fault lifted a section of the reef high into the air, brandishing the dark rock for all to see. The trail was steep—a gauntlet of narrow switchback turns, full of slippery boulders that tested one's stamina and balance. Yet the site was still quite popular, particularly with geologists and paleontologists, for it led into the fossil remains of an ancient world.

Park geologist Hearst was the keeper of this treasure, and was astute and knowledgeable about its intricate secrets. But she was as exuberant as she was scholarly. She told me she had last hiked up this reef just two weeks earlier, yet she wore a big smile despite some heavy breathing. "It's a geological Disneyland," she proclaimed. "Every time I go up there, I learn something new. How many times have I been on that ride, you ask?" She shook her finger toward the reef. "Don't know, but I want to do it again!"

The hike began at the bottom of an enormous depression known as the Delaware Basin, which spread out into Texas. The reef had formed over a distance stretching many miles around the lip of the basin, the horseshoe mouth of which once pointed out to an ancient sea. A quarter of a billion years ago, this reef was still glowing with a halo of life formed by millions of juvenile fish and other marine creatures that once used the nooks and crevices of the reef to avoid large predators.

Back then, two enormous continents—Laurentia (made up of North America, Europe, and Asia) and Gondwana (made up of South America, Africa, Antarctica, and Australia)—formed the terrestrial landscape at the surface of the planet. These two landmasses were on a collision course, soon to form Pangaea, the single continent that would take the world through the Permian extinction, an event that came the closest to ending life on earth than at any other time in the last 600 million years.

Our trail told the grand story of life before that event. We scrambled up the loose rock beneath the slopes of the giant reef head. We gained altitude quickly as the trail rose above the desert landscape. This was a deepwater reef, different from the shallower coral reefs most recreational divers are familiar with today. In Permian times, we would have been walking 5,000 feet (1,500 meters) below the surface of the ocean. "A very long snorkel to get to the top," said Hearst.

As we rose upward, larger boulders and layered outcrops gradually displaced the loose, rocky slopes. Hearst stopped before a large boulder that reached our height and stared at the markings on it. At first I didn't see anything special; it was just a big boulder. But then she pointed out the many fossils contained in the rock. It turned out that we weren't staring at a plain rock—we were gazing on the calcified remains of ancient reef animals that had once been bound together in a mass of life.

During the Permian period this gallery of life included flowerlike crinoids, which sat atop stalks attached to the seafloor, their numerous tentacles coated with mucus extended out to capture prey, and you could see the fossil remains of these creatures in this rock. There were also bryozoans, small animals that superficially resembled corals, which grew in tightly packed colonies resembling intricate fans, lacy fronds, or fruitlike displays that accumulated into massive stony buttresses. Also here were clamlike creatures called brachiopods, which were filled with a tangle of filaments that helped the animal sift food from the water but which would have made a poor clam chowder. There were numerous species of sponges as well as nautilus-like creatures housed in large spiral shells. The boulder was filled with such animals, surrounded by algae, which acted as cement to hold everything together. As she pointed to other rocks nearby, my astonishment grew. All the boulders housed similar amazing displays.

From the base of the reef we pressed on up the trail. As we approached the part of the reef formerly within the reach of sunlight and the energy of the waves, the reef fauna began to change from marine communities dominated by sponges and bryozoans to those dominated by algae and large clamlike gastropods.

Toward what was once sea level, the sponges disappeared. We entered the intertidal zone where outgoing tides would have periodically exposed the reef to sunlight and air, and this produced still more shifts in the animal communities. Ahead we could see the remains of limestone barrier islands. Behind the barrier islands were sand and gravel bars cut through by tidal channels, and beyond that the dry remains of a large lagoon facing a shoreline of salt flats.

The Permian period stretched from 298 million to 251 million years ago, the reef thriving across the West Texas terrain along the margins of what was once a warm tropical sea. In its prime, it would have been about four hundred miles in length.

Reefs are among the most biologically diverse of any ecosystems. They are the rain forests of the sea. Yet they leave more evidence than a rain forest for the paleontologist to study because they are made up of hard-bodied organisms that make fine fossils. It's why paleontologists have made the pilgrimage to McKittrick Canyon for decades to witness what nature has exhumed almost intact.

* * * *

It hasn't been that long since man would have looked at this towering monument to the history of life and not understood what he was seeing. The recognition and study of fossils in rocks grew out of an incident in the late fifteenth century when two fishermen caught a giant shark off the coast of Livorno, Italy. The local duke sent the shark to Niels Stensen (aka Nicolas Steno), a Danish anatomist working in Florence. Steno dissected the animal and noted how much the shark's teeth looked like "tongue stones," triangular pieces that rock collectors had been gathering for ages. Few at the time would have conjectured that tongue stones or any other fossils might be remnants of ancient sea life, but Steno started making a case for it and was widely credited with giving birth to the science of paleontology.

The awareness of fossils grew, and in 1815, William Smith, a geologist from the county of Oxford, England, published a complete geological map of England and Wales. He was the first to use fossils as a tool for dating and mapping rocks by their stratigraphy, the lines and layered elements of earth that are visible when sedentary rocks are cut into—though it wasn't until after Darwin that scientists realized the importance of these fossils to understanding the timing of evolution.

Geologists discovered that layers of rock in North America could correspond in time to layers of rock in Asia or even Africa and that similarities in the fossils within them could be used to determine their synchronicity. But what geologists began to realize was that the layered record of earth's history at times told the story of evolution a bit differently from Darwin. The master believed that evolution advanced in tiny increments over multiple generations and that the process was geologically slow. *Natura non facit saltum* ("Nature makes no leap") was his credo. But other scientists began to note a number of upheavals captured in the rock record of earth's history, which showed radical, sudden changes in animal fossils.

These upheavals presented an amended look at Darwin's grand scheme, and were known as mass extinctions. Evolution continued after them, but mass extinctions reordered nature, abruptly ushering out older forms of life and allowing for the creation of newer ones.

Simple animals without shells or skeletons appeared about 635 million years ago during the Ediacaran period, when oxygen in the atmosphere began to build toward present levels. Since then, there have been five mass extinctions. Evidence of the Permian period, which preceded the Permian extinction 252 million years ago, surrounded National Park Service geologist Hearst and me.

Perhaps the most famous of the five extinction events was the one that wiped out the dinosaurs at the end of the Cretaceous period about 65 million years ago. Scientists long argued over what had killed off the dinosaurs until, in the late 1970s, a team of scientists at the University of California, Berkeley, came up with a theory. Luis Walter Alvarez, a bespectacled Noble Prize–winning nuclear physicist and leader of the team, found unusually high levels of iridium—a heavy substance rarely found on the surface of the planet, but quite common in meteorites—in layered deposits of earth that represented the Cretaceous extinction in both Italy and Demark.

Alvarez, his son the geologist Walter Alvarez, and colleagues shook the scientific community with their announcement that the mystery of the Cretaceous extinction had been solved: an asteroid got the dinosaurs.

Scientists were at first skeptical. Older hypotheses cited volcanism or glaciation as the primary cause of this mass extinction. But eventually high levels of iridium were found at more than one hundred sites, all marking the Cretaceous extinction, and the evidence couldn't be ignored. But where was the crater?

The Alvarez team went looking for a depression somewhere on the planet big enough to have fit the job. The team calculated that the asteroid must have been about seven miles in diameter. In June 1990, a decade after the original Alvarez proclamation, geologists discovered a huge crater underlying the northern tip of the Yucatán Peninsula near the town of Chicxulub ("Chick-sha-loob"), Mexico, from which the crater eventually took its name.

The crater revealed that the asteroid must have been about 7.5

miles (12 kilometers) wide and was traveling about 44,640 miles per hour (20 kilometers per second) on impact, roughly twenty times the speed of a bullet. The collision would have released a million times more energy than the largest nuclear bomb ever tested.

The impact blasted thousands of tons of rock as well as the mass of the asteroid back into the atmosphere, with some elements going into orbit, while others returned to the ground in a barrage of flaming meteors. These fireballs ignited the verdant late Cretaceous landscape, burning half the earth's vegetation in the weeks following the impact. Dust along with the smoke from the fires obscured the light of the sun, dealing a deadly blow to plant life.

In the ocean, huge tidal waves spread out to the continental shores, leaving a line of beached and bloated dinosaurs skewered on shoreline trees. Scavengers had a field day on the plentiful carcasses. After the initial fires burned out, the earth descended into a period of perpetual night caused by a blanket of smoke and dust in the air. Trees and shrubs began to die, as did the animals that ate them and the carnivores that ate the plant eaters. The Cretaceous extinction killed off the dinosaurs and many but not all of the mammals.

At the top of the Capitan Reef, we looked out over the fossils, rocks, precipices, and the valley below us, and imagined life over 250 million years ago at the pinnacle of the Permian period. Dry land, which was then about fifteen miles northwest of the reef, was growing drier. The lush swamp forests that had existed before the Permian had been replaced by conifers, seed ferns, and other types of vegetation that were drought-tolerant. Giant cattail-like trees grew up to eighty feet. Ten-foot relatives of the centipede splashed through inshore water.

The first vertebrates had crawled out onto the land only about 100 million years earlier. Giant amphibians, which roamed the marshlands, were up to six feet in length and two hundred pounds in weight. They sucked down dinner with enormous mouths filled with sharp teeth, tossing their captives little by little back into their deep throats, like a crocodile or alligator would. There were flying lizards and large armored herbivores the size of oxen. There were a number of sharks in the Permian oceans, the most bizarre being *Helicoprion*, which had a spiral jaw fitted with backward-leaning teeth that looked like a buzz saw. Primitive pelycosaurs about ten feet (three meters) long with smooth bodies spread over much of the land with giant swordfish-like fins on their back for capturing the sun.

The Permian world was a lively one, as proven by the numerous fossils that adorn the earthen walls of McKittrick Canyon. But something caused the annihilation of most of these animals.

THE SECOND CREATION OF LIFE

The Capitan Reef that decorates the top of the Guadalupe Mountains above McKittrick Canyon is similar to the structure of Mount Rushmore, only carved not with US presidents but with the force of life that thrived before the mass extinction. Yet the rocks in McKittrick Canyon do not display evidence of the end of the Permian.

To see that, Sam Bowring, a bearded and amiable professor of geology whom I visited earlier at MIT, had to travel to China. Bowring showed me a photo of himself and Zhu Zhuli, a Chinese researcher, in Meishan, standing on the face of a rock quarry. Zhuli had his feet on a dark line in the rock that represented the end of the Permian. The change in color was caused by a dramatic change in the geology and chemistry of the rock. It was the geological boundary line between the Permian and the Triassic periods, the point where one era of life encased in sediments of earth ceased to exist and another was laid down on top of it. In the photo, Bowring stood above the line in early Triassic ash beds. It is one of the best-studied Permian-Triassic boundary sequences in the world. Fully 333 species have been identified in the fossils below where these two scientists were perched. But above that line almost all of them disappear, an extinction rate of 94 percent.

John Phillips, a mid-nineteenth-century English geologist who published the first global geological time scale, found that the fossils were so different on either side of the Permian-Triassic boundary that he referred to the line in the stratigraphic layers that Bowring stood above and the difference in fossils on either side as the Second Creation of Life. He never saw the line in Meishan, China, but had studied this event at similar stratigraphic sites elsewhere in the world.

The catastrophe that created this boundary has similarities to the destruction humans are inflicting through greenhouse gas buildup, ocean acidification, and global warming. No, it wasn't a giant spectacular meteor falling out of the sky. The primary villain of the Permian extinction was the Siberian Traps. This eruption occurred about 252 million years ago, according to new findings from Bowring. At that time a viscous magma flowed out of the ground and spread over the land, filling in the valleys and basins around it like honey finds the crevices on a piece of toast. The total amount of lava flow was mind-boggling. In one area it grew 6,500 meters thick, almost four miles. "In the end it covered much of Siberia, an area close to the size of the continental United States," Bowring told me.

Still, there was not just a single cause to this extinction. It was more the perfect storm, the coming together of multiple perpetrators, as it has been with other extinction events. The lava that created the traps burned up through an enormous coal reserve at its center, and the heat of the molten lava converted much of the black rock to CO_2. But as temperatures rose, some of that coal would have converted to methane, which is twenty times more potent a greenhouse gas than CO_2, and this would have accelerated warming.

The end result of the buildup of CO_2 and methane, among other causes, was one of the few mass extinctions of insects in earth's history. Their numbers descended from sixty families during the height of the Permian period to almost zero at the end of it. The air was silent, since birds had yet to evolve. The coal that had thrived in the marshy environments and plentiful vegetation disappeared as the earth grew drier. Whole forests and entire ecosystems of plants died but fungi flourished, since they fed off the dead plant and animal matter.

Though the asteroid that got the dinosaurs at the Cretaceous extinction may have produced a better fireworks display and spectacular tsunamis, in terms of pure raw killing power, the Permian extinction can't be beat. Its witch's brew of toxins poisoned the land for several hundred thousand years. Doug Erwin says that the eruption of the Siberian Traps caused global cooling from the erupted dust, global warming from the CO_2, and acid rains from billowing clouds of sulfur. Couple this with ocean acidification and the death of oxygen in the deep seas due to the melting of polar ice and the loss of ocean currents, and you have a lethal force that far exceeded the destruction caused by the falling asteroid during the Cretaceous.

The resulting excess CO_2 entered the ocean, making the water acidic enough to prevent animals from forming exterior skeletons and destroying most of the reef-making organisms of the Permian seas and most of the reefs. The acidic nature of the seawater, coupled with the lack of oxygen in the deep oceans, wreaked havoc on marine plants and animals. The sulfates that emerged from the volcanoes reached the upper atmosphere, to be carried afar as sulfuric acid and lethal acid rains. These rains may have been strong enough, suggests Erwin, to kill off many of the terrestrial plants. This totally denuded landscapes over much of the earth's surface. Scientists have found evidence that much of the rain that followed the Permian extinctions rolled off the land in flash floods, since there was no vegetation to contain the flow of water.

Floods skipped across the earth like oil does on a hot skillet, moving in every direction, leaving braided gullies in the rock record. I've witnessed fast-moving desert flash floods that carved out chunks of road like butter, but desert rains are meager. Imagine flash floods raging in plant-free tropical or coastal environments where annual rainfalls are twenty, fifty, one hundred inches, or more, racing in full and furious force across landscapes stripped of vegetation, and you'll get an idea of what the floods that followed the Permian extinction must have been like.

But despite the evidence of multiple causes for the Permian extinction, some scientists still champion their favorite antagonists. Andrew Knoll, a paleontologist at Harvard, thinks that many of the catastrophes—their causes and their results—can be boiled down to one chemical compound, CO_2, the biggest villain of the day, and perhaps our greatest threat as well. In a 2007 paper in *Earth and Planetary Science Letters*, Knoll and colleagues tried to work backwards from the extinction event, doing a computerized autopsy of the victims to see if the massacre matched the typical scenario caused by oxygen depletion, a breakdown of the food web, and acid rains, but none of them quite matched the autopsy except for CO_2. He highlighted a gas that so many today ignore. "Only 30 percent of the species of plants and animals were tolerant of massive doses of CO_2. But after the Permian extinction, that 30 percent suddenly becomes 90 percent of all living animals."

Estimates vary on how long this extinction lasted. MIT's Sam Bowring sets the duration at about sixty thousand years. The tiny chewing apparatuses of small eel-like animals are some of the first fossils to appear in layers of earth laid down after the Permian extinction. Fossils of *Lystrosaurus*, a mammal-like reptile that looked like a bulldog with tusks, but which survived the extinction and proliferated, mark the beginning of the Triassic recovery.

The irony of the Permian extinction is that though it devastated large portions of the planet, it created opportunities in newly emptied terrain. From the resurrection of life after the mass annihilation of the Permian came more-adaptive species, changes in ecosystems, and a world more diverse than the one before it. Perhaps these improvements could be in our future—if we survive the extinction.

The processes were similar to what Darwin witnessed in the Galápagos Islands. Of the twelve species of finches he collected, all were adapted from a few individuals from the mainland or other islands, which had arrived in the Galapágos and proliferated after finding no competition for the banquet of seeds available.

What emerged from the Permian extinction was a similar explosion of new animals and plants. Life not only survived, it eventually thrived. By 225 million years ago, the first dinosaurs appeared; but by 65 million years ago the group, other than birds, was gone. Their reign on earth lasted close to 160 million years, a length of time that the family of man has barely approached.

Though the end was glorious, the millions of years it took to recover from the Permian extinction were excruciating.

After a brief lunch at the end of the trail, looking out over western Texas and southeastern New Mexico, enjoying the cooler breezes at the top of the range, Hearst and I gathered our gear and headed back down the same path we'd come up on, still taking note of the various changes in the fossil communities, enjoying a second look, knowing them better.

Hearst explained that life as a whole eventually resurrected itself from the Permian extinction, but few of the individual species of plants and animals displayed here in the fossils of the mid-Permian made it across the boundary. "Life goes on. Life is incredibly resilient. But my work here has taught me that ecosystems and individual species are so very, very fragile," she said. If history is our teacher, then life in the aftermath of our own era will prove equally resilient, though right now ecosystems and individual species are rapidly disappearing.

From the height of the trail we looked out over the vast desert below and reflected on our own situation. We stood in the middle of evidence of a past evolutionary catastrophe and gazed out over another in progress: our own. Some scientists believe our current situation started at the onset of the Industrial Revolution in Great Britain during the 1700s. This is when CO_2 in the atmosphere began its upward climb, a change mirrored here in the aftermath of the Permian. But others date the commencement of our dilemma to 1800, when the human population reached one billion.

Still, others say we entered the present biodiversity crisis during the final moments of the last ice age from about fifteen thousand to twelve thousand years ago, when a substantial portion of the large animals that once existed in North and South America disappeared. Similar scenarios took place in Australia, New Zealand, Europe, and Asia with the arrival of man.

Hearst poured some water on a group of fossils, washing off the dust and making them more defined and lifelike for the moment. Of course, the evolutionary processes that produced their first spark of life were much more complicated.

2

ORIGINAL SYNERGY

IT HAD BEEN RAINING off and on all week at the Cary Institute of Ecosystem Studies in Millbrook, New York, a reserve consisting of two thousand luscious acres of mid–Hudson Valley oak, maple, and hemlock that the institute refers to as its "campus." I had some big questions to ask—how life got started, how evolution drove its development, what role oxygen played, if evolution was still at work in the natural world—and I began my quest at Cary.

A heavy mist rose that day from the wetland patches and crept through the forest propelled by a Sunday morning sun. William H. Schlesinger, biogeochemist and president emeritus of the institute, and his wife, Lisa Dellwo, guided me on a birding expedition during a break in the rain through the woods and meadows that abound there. We spotted seventy-six birds before breakfast, sixteen species in all. When I couldn't see a bird, they both went to extremes to describe the bird and the place in the woods where it was. Lisa claimed birding cultivates cooperation and communication, and is sadly overlooked as executive training.

On the way back from the woods, I got to talk to Schlesinger about life. Bill is a tall, burly man with a hearty laugh, a deep, articulate voice, and a head full of chemical formulas. He thinks chemistry is often underrated and coauthored a book called *Biogeochemistry: An Analysis of Global Change* with Duke University's Emily S. Bernhardt. The book looks at the role of biology, geology, and chemistry in changes that have occurred on earth.

"The road to life on planet Earth was peppered with more chemistry than people give it credit for," Schlesinger told me.

Though our galaxy, the Milky Way, has existed for 13.7 billion years, our solar system is only about 4.6 billion years old. Our sun, said Bill, is at least a second-generation star—a descendant from a prior supernova, a large star that ran through its nuclear fuel, collapsed, and then exploded. That explosion blew a whole lot of dust and particles into the cosmos, and the sun and Earth coalesced out of that cosmic residue. A heavy meteor bombardment ensued during the first billion years, added to Earth's mass, and created its moon. The heat of the collisions and the radioactive decay of the materials melted the whole ball with the heavier chemicals sinking to the molten core, while the lighter elements formed the semifluid mantle and the crust that floats upon it.

One of the critical components for life, Bill pointed out to me, was plenty of water. At the Cary Institute, it had just rained, and we jumped around puddles and dodged the occasional deluge delivered from the leaf canopy above while Schlesinger explained how we got all this moisture. Schlesinger is an excellent orator and teacher, one who is not afraid to hold forth until he sees the light in your eyes that tells him you got it.

Water probably came from the same bombardment of materials that built the planet, he suggested. The heat of the planet would have kept that fallen moisture as steam in the atmosphere until the Earth's surface cooled to 212 degrees Fahrenheit (100 degrees Celsius), the boiling point of water. Afterward, the steam coalesced and the moisture fell out of the sky over several million years, filling the oceans.

The sun was then 30 percent less luminous than it is today, but the presence of water vapor and CO_2 in the atmosphere produced a greenhouse effect, catching any escaping infrared or heat radiation and redirecting it back toward the surface of the Earth. This warmed the planet. Without the greenhouse effect, the Earth today would be mostly covered with ice and have an average temperature of about 0 degrees Fahrenheit (minus 18 degrees Celsius).

Another gift of the early arrival of celestial materials on Earth was carbon, a critical element of life. "All life on this planet is made of compounds that have carbon in them," said Schlesinger. Carbon forms strong bonds with other chemicals, which is important for building long chemical structures like proteins, cellulose, and DNA. "If you took your body, dried all the water out of it, what's left would be about 50 percent carbon," he said. "We are basically bags of carbon running around on the surface of the Earth."

How did we get life from carbon molecules? Where did it first occur? These have not been easy questions to answer. Some interplanetary dust and comet ices are found to contain organic matter containing carbon and could have survived entry into Earth's atmosphere, adding to the carbon already here, he says. Even if the total amount of organic comet matter received by Earth was small, these elements could have served as a catalyst for life.

Scientists and philosophers have debated the question of first life for millennia, though most of the explanations have centered on fable or religion. In 1929, British biochemist J. B. S. Haldane and Soviet scientist Alexander Oparin independently suggested that all the ingredients for life existed on Earth from the beginning and that energy from the sun and some unknown process had gotten life started. In the 1950s, Stanley Miller, a doctoral student in the laboratory of Harold Urey, at the University of Chicago, got more specific when he attempted a famous experiment. He mixed ammonia, methane, and hydrogen—a commonly accepted recipe for the early atmosphere and ocean—in a big laboratory flask and subjected it to an electric charge that simulated lightning. He analyzed samples at regular intervals. The result was a jackpot for Miller and the Urey lab: after about a week he found simple organic molecules in the flask. Life could be produced in a laboratory. He had cooked the infamous primordial soup.

But fame was fleeting. Miller had fashioned his recipe after Jupiter and some of the outer planets, but those models proved to be an inaccurate representation of early Earth. More realistic versions didn't do as well, either, and the idea that you could cook up life like soup fell out of favor.

But if soup didn't initiate life, then what did? Scientists turned to the oceans for answers.

Possible solutions emerged in the early 1970s when scientists noticed rising plumes of warmish water along a deep ocean crack near the Galápagos Islands, the same islands that fostered Darwin's theory of evolution. In 1977, the US naval submersible *Alvin* dove down 7,000 feet (2,100 meters) to investigate deep-ocean geysers and found a wonderland of giant clams and mussels as well as eight-foot-long tubeworms. The sheer abundance of life at that depth was astonishing—a tropical rain forest of ocean species. Here, eyeless shrimp and snails munched on mats of bacteria that thrived on sulfur compounds. These underwater geysers generated supplies of energy for the plants and animals, rather than the sun, whose light didn't reach this depth.

Scientists have since explored over two hundred of these geyser systems in the oceans. Some of the most remarkable have been along the deep-ocean ridges of the Pacific, Atlantic, and Indian Oceans. At these ridges, the seafloor spreads outward along a rift fed by hot magma below, the birthplace of new land on earth. At such places, researchers found colossal deepwater chimneys known as black smokers, some as tall as skyscrapers, pumping what appeared to be billowing black smoke into the sea. It wasn't real smoke, of course, but boiling metal sulfides welling up from the magma below, the acidic mixture oozing into the water at 662 degrees Fahrenheit (350 degrees Celsius).

Was this eerie place with its bizarre cast of characters where life originated? Though boiling sulfides hardly sounded like a Sunday buffet, there were certain advantages. The ocean depths would have shielded life from the UV radiation that was then pummeling the ocean surface as well as the land a few billion years ago. Michael Russell at NASA's Jet Propulsion Laboratory in Pasadena, California, thought the mixture too acidic to be involved. So he came up with a theory for a milder first-life solution by looking for another type of geyser that had a gentler origin.

His theoretical answer arose from the slower movement of fresh crust across the seafloor, exposing rocks from the mantle. Russell's candidate for the prime mover wasn't an acidic mixture of superheated waters; it was the reaction of freshly exposed rock with seawater at a relatively cooler 210 degrees Fahrenheit (100 degrees Celsius).

Seawater expanded the rock, creating fissures and cracks, which drew in still more seawater. This process released energy as heat and large amounts of hydrogen and methane gas. This created another type of hydrothermal geyser, which some called white smokers, or more accurately alkaline vents. Rather than creating a black chimney with a single orifice belching black superheated smoke, these vents were complex structures with mazes of tiny compartments that exuded warm alkaline water to the surrounding cold seawater.

Life could have arisen from sulfidic submarine hot springs situated some distance from the deep oceanic ridges. Scientists thought that four billion years ago life could have emerged there from a mass of bubbles, each bubble containing hot mineral-laden solutions.

Around the turn of the twenty-first century, the research vessel *Atlantis* and its human-occupied submersible *Alvin* found this exact type of geyser about nine miles (fifteen kilometers) from the Mid-Atlantic Ridge. Dubbed the Lost City, these vents stood like ornate structures up to two hundred feet (sixty meters) in height, poking up into the vast darkness. At this depth hydrogen could more freely bind to carbon dioxide to form organic molecules. First life was not a single cell but a rocky labyrinth of mineral cells that produced complex molecules, including the formation of proteins and eventually DNA molecules, generated by the energy of the warm vent fluids.

As we came to the end of our bird walk at Cary, Schlesinger said that this made sense. He had one caveat: he favored a more neutral solution for first life. "Life can tolerate a wide range of pH, but really acid conditions [low pH] are likely to oxidize organic materials and really alkaline conditions break down cell membranes," he explained.

OXYGEN MAKES IT HAPPEN

Most scientists agree that, for the first few billion years, life was largely microbial. Yet these little critters were responsible for most of the genetic heavy lifting. Though we marvel at the size and anatomical complexity of large animals, these features were made possible by cell biology and genetics that were developed in single-cell creatures in much earlier times. According to Harvard's Andy Knoll, when complex life first evolved, it had the majority of its DNA already worked out.

For life to really get going, to produce the complex forms of more evolved beings, it had to have oxygen. Two and a half billion years ago, "life" was still in bacterial forms. It had its genetic architecture, but it survived in oxygen-free environments, so it stayed small. But then some of the oxygen-free bacteria evolved into cyanobacteria or blue-green algae, the stuff you sometimes see on polluted waters, commonly referred to as "pond scum."

These guys promoted photosynthesis, a different type of metabolism from what their archaic brethren employed. Photosynthesis used sunshine, water, and carbon dioxide to produce carbohydrates and, finally, oxygen. At last, the giraffes and basketball players of the world had a chance at survival!

Oxygen was the critical element in the burst of evolution that occurred during the Cambrian explosion about 570 to 530 million years ago, when most of the major animal groups suddenly appear in the fossil record. At the time the air was murky, since there wasn't enough oxygen to scrub the atmosphere of haze and dust. Without enough oxygen, there was no ozone, either, so the searing intensity of ultraviolet light from the sun could fall without obstruction. Ultraviolet light breaks up water (H_2O), and since hydrogen (H) is so light, it can slip into space, and there goes your ocean. Without oxygen holding on to hydrogen, the world today might look a lot like Mars: a dry, dusty, pockmarked planet with no seas, lakes, rivers, or streams and no visible sign of life.

Oxygen gradually accumulated on earth from the photosynthesis of plants. Once oxygen reached critical mass, changes were sudden. If you look at the paleontological record in the soil, there is evidence of oxygen-free microbes in one layer, followed closely by oxygen-dependent microbes in another layer. This introduction of oxygen, though a boon to most life, spelled destruction for a good deal of earth's early ancestors who excelled without it.

Oxygen made the planet livable. Once established, oxygen patrolled the atmosphere capturing all the hydrogen atoms trying to get away and turned them back into water and rain. Now an ozone shield could form, dampening the intensity of ultraviolet light. All plants and animals depend on oxygen as part of their life cycles, lonely exceptions being the microscopic nematode worms that get along in the stagnant oxygen-free depths of the Black Sea and the creatures that survive on those deep-ocean geysers.

The beginning of animal life had to wait about four billion years until atmospheric oxygen began to rise toward present levels. According to Andrew Knoll, complex multicellular organisms and oxygen first appeared in the fossil record some 580 million to 560 million years ago, during the Ediacaran period. "The oxygen increase pushed earth toward its present state, but it didn't achieve it all in one go," he told me when I visited with him at Harvard.

Life didn't burst forth onto center stage in full and varied forms until the Cambrian period from 542 million to 488 million years ago.

The Burgess Shale, the famous quarry of Cambrian life discovered in 1909 by Charles Walcott, a paleontologist and former director of the Smithsonian Institution, sits high in the Canadian Rockies on the eastern border of British Columbia. It is perched at about eight thousand feet on the western slope of a ridge connecting Mount Field and Mount Wapta in Yoho National Park, near the tourist destinations of Banff and Lake Louise. The view from the rocky slopes of the Walcott Quarry—surrounded by a thick conifer forest, Emerald Lake below, and the snowcapped Canadian Rockies beyond—is one of the finest on the continent. Walcott's daughter, Helen, wrote to her brother Benjamin in March 1912 when she was touring Europe, describing castles, fortresses, the Appian Way, and the Roman aqueducts, "but I'd prefer Burgess Pass to anything I've seen yet," she said.

Our first really good display of what nature was up to during the Cambrian explosion didn't materialize until 530 million years ago, when mudslides at the Walcott Quarry captured a broad selection of fossil samples reflecting much of the Cambrian's incredible animal diversity. Among paleontologists, the stature of these finds can only be appreciated when you take into consideration that since this geological period, and over a vast range of time in which life has been through enormous changes and upheavals, no new body designs, no new phyla, have been added to the collection of life displayed at the Burgess.

The Burgess Shale is a miracle of preservation. Stephen Jay Gould proclaimed in his book *Wonderful Life: The Burgess Shale and the Nature of History* that mammalian evolution "is a tale told by teeth mating to produce slightly altered descendant teeth." Which is to say that if it weren't for teeth, we wouldn't know as much about our ancestors. Teeth outlast everything else and are the dominant feature in any anthropological collection.

But soft body parts like stomachs and other fleshy bodily organs and appendages in the wild collection of creatures found at the Burgess Shale? You really have to be lucky to get samples of any of those from the distant past. About 20 percent of the 140 or so original species found in the Burgess Shale were skeletonized, and the rest were soft-bodied. Still the earth that captured these creatures clearly displayed their ghostly impressions. This incredible find is preserved in a section of the shale about the height of a man and not quite as long as a city block, and according to Gould it has "more anatomical disparity than in all the world's seas today."

In a burst of evolutionary creativity, all the major body plans suddenly appeared onstage. Although some scientists wonder if the original cast was so varied, Richard Leakey argues that as many as seventy actors were present, displaying the different body plans or phyla of life. But what remains today are perhaps only thirty or so such plans, the others having been cut from evolution's cast since the Burgess Shale was formed.

The Smithsonian's Charles Walcott was of a more conservative opinion. He originally categorized all the creatures he found in the Burgess Shale as a part of the recognized phyla or body plans of today. But in the late 1960s, Harry Blackmore Whittington, a paleontologist at Cambridge University, reopened Walcott's excavation to take another look. As at the Capitan Reef at Guadalupe National Park, these glorious remnants of past life were entombed in the crest of a mountain, but they had once inhabited an ancient sea. The residents of this ocean community had been caught by mudslides, which preserved their bodies in ghostly detail as flattened images in thin layers of shale.

They were an odd bunch, mostly small but truly varied and exotic: *Opabinia* had five eyes and a long, flexible trunk tipped with a grasping spine. *Amiskwia* looked like a strange seal with a rattlesnake's head. *Anomalocaris* had underwater wings, shrimp tails for arms, and a scary mouth with a ring of sharp teeth for cracking the bodies of scorpions, spiders, and shrimp. *Wiwaxia* had a series of spines projected in two rows along its back, looking like a bear trap ready to be sprung. And last but not least, there was *Pikaia*, a worm an inch and a half long—man's early ancestor.

The Burgess is our best example of the Cambrian explosion, a period during which life jumped from a simple and not too varied existence to the ancestors of the fullest complexity of nature we've seen on the planet. The Cambrian explosion perplexed Charles Darwin because it offered another refutation of his conception of evolution as a slow and steady progression. Here, life quite suddenly made an enormous leap.

The development of vision during the Cambrian was one of the great inventions of nature that may have helped ignite the Cambrian explosion, transforming the behavior of all living creatures. It put prey at a whole new level of desperation. Predators could better spot and chase prey. This led to the evolution of shells and the tough exterior skeletons of crustaceans, giving prey a chance at survival. It also provided a much greater likelihood that these creatures would appear as fossils in the rock record because those tough exteriors survived time.

Movement was another one of nature's great inventions, but appears to have shifted gears in importance after the Permian mass extinction 250 million years ago. Life in the Permian oceans was largely anchored to the bottom. Lampshells, sea lilies, and shellfish filtered food from the water, a meager though lazy way to make a living. But after the Permian extinction, things that moved dominated the animal kingdom. This new skill buffered life from sudden change in the environment, allowing it to develop.

But another important aspect was that movement led to complexity. Nature was more diversified after the Permian. Rather than a handful of species that dominated the landscape, with the rest left to eke out a living, multiple species began to abound and thrive in conjunction with each other. The number of species living together increased dramatically in the fossil record and laid the foundation for the world we live in today.

AFRICAN SOJOURN

Animal life has grown quite larger and more complex since the Cambrian explosion. To see evolution at work I visited the Ngorongoro Conservation Area in Tanzania, Africa. Man has devastated large animal populations in most places on earth, but in Africa these two evolved simultaneously and wildlife adapted to keep their distance. Nowhere on earth except in Africa are there so many large animals, though even here they are subject to human voracity. Evolution, however, is helping some animals adapt to man by shedding their tusks and horns.

To view this firsthand I traveled one summer day with Joseph Masoy, a smiling, husky Tanzanian, who commanded his Toyota Land Cruiser over bumpy African roads to get into Ngorongoro Crater, the relic of an ancient volcano that was once filled with lava but is now filled with African wildlife. Seated with me in the truck were Nicholas Toth, Kathy Schick, and James Brophy, all professors at Indiana University, who were traveling to Olduvai Gorge. The car contained several weeks' worth of gear, supplies, and personal belongings, as well as a pop-up roof that allowed us to view and take pictures of the wildlife along the way without getting eaten.

We gradually approached the green jungle that shrouded the Crater Highlands of the Eastern Rift Valley, as the sun boiled up the midday tropical clouds into the sky. By early afternoon we crested the rim of Ngorongoro Crater and descended into the ancient cauldron. Ngorongoro Crater became a UNESCO World Heritage Site in 1979. At our first look into the crater, it seemed vacant: some little specks down there—rocks perhaps—but not much wildlife.

As we wound down the inside wall of the cauldron, these specks grew more and more spectacular. The first group we came upon turned out to be a herd of Cape buffalo. For the most part the animals ignored the tourist vehicles. They moved about in small groups and herds, feasting on the savanna grasses that covered the crater's floor. One buffalo stood and stared at our truck. It appeared mean and perturbed. Masoy claimed that buffalo are some of the most dangerous of Africa's wildlife, partly because there are so many of them and partly because people don't take them seriously. I could only count them in batches, each containing perhaps fifty animals. There are at least twenty other batches within our field of view, perhaps one thousand animals in all.

We spot a pair of rhinos. They keep their distance, maybe two hundred yards away. That afternoon we spot about fifteen hundred zebras, two thousand wildebeest, one thousand buffalo, several bustards (a large terrestrial bird), black-crowned cranes, impalas, six hyenas, about eight jackals, one African lion, one cheetah, and eight giraffes.

But the surprise of the day came when we spotted three elephants in Ngorongoro Crater walking through a crowd of several hundred zebra. One elephant was tuskless; another seemed to have broken one of its tusks. None had a glorious pair of ivory as in your typical African photo. This was evolution in action. The tusks of the big elephants are a gold mine and too dangerous for the animals to carry.

Despite the government threat to shoot poachers on sight, poachers keep trying. Similar to other parts of the world, Africa is losing its animals. Neither strict national laws nor international support nor tourist income completely protects these majestic animals from illegal hunters. Congolese authorities recently accused the Ugandan military of killing twenty-two elephants from a helicopter and then carting away more than a million dollars' worth of ivory.

In the 1970s, 10 to 20 percent of all the elephants in the wild were killed. At that rate, extinction could have come quickly, but international pressure and evolution have given the elephants a reprieve. Poaching put evolutionary pressure on animals with tusks, and tusks on elephants began to disappear rapidly. Ownership of ivory tusks was too expensive.

Selection has affected both male and female elephants. A Princeton ecologist, Andrew Dobson, traced the evolution of tusklessness in females at five African wildlife preserves. In one park where elephants were relatively safe, the incidence of tusklessness in females was small, a few percent. But in another park where they had been heavily poached, it was a different story. Females aged five through ten were about 10 percent tuskless. But females aged thirty to thirty-five were about 50 percent tuskless.

Researchers have noted similar results for males. That nature would allow male elephants to give up their tusks is phenomenal.

Males use their tusks to battle each other for access to the females. A male without tusks is like a knight without a lance; yet, due to the state of game hunting, tuskless males have a better chance of surviving. Thus nature now selects for tuskless males as well as females.

Adapting to man is currently wildlife's greatest evolutionary challenge. The animals in Ngorongoro Crater, including the human ones in the safari vehicles, are all descendants of our common ancestor *Pikaia*. Yet we are presently locked in mortal combat.

The diversity of life is present in Africa's game preserves, but one wonders how it began and how long it can continue.

3

THE GROUND BELOW THE THEORIES

MOST SCIENTISTS AGREE that the slow, lateral movement of the continents, their joining and separating on the surface of the planet, has strongly influenced the broad diversity of plants and animals that exist on earth today. During the Permian period all the continents joined together in one enormous landmass, a super-continent, but after that extinction event, the supercontinent Pangaea began to split apart like a broken dinner plate with its pieces scattered across the oceans. This separation of the landmasses led to a corre-sponding separation of species of plants and animals. Newly separated species no longer exchanged genes with one another and over time isolated populations evolved away from each other and became sepa-rate species.

The splendor of this evolutionary tale was on display when Dar-win and the crew of HMS *Beagle* rowed up to the island of San Cris-tóbal, a black mound of volcanic rock in the middle of the Pacific Ocean, during their celebrated visit to the Galápagos Islands that began on September 17, 1835. At a distance the island looked deso-late, but upon landing Darwin found it covered with plants that bore leaves, flowers, and seed-bearing fruits. He had been on a four-year expedition to South America and was now heading across the Pacific on the long way back home to England. The crew of the boat hoped to catch a tortoise for some meat to make a tasty soup, but there were no tortoises on San Cristóbal.

He did find numerous birds, which were unfazed by human presence as they looked for seeds in the bushes; they'd had no experience with humans. One of the crew even caught a bird with his hat. Darwin picked up an iguana and threw the animal into the water over and over, but each time it swam straight back to him. He yanked on the tail of another that was digging a burrow, and it turned and looked at him as if to say, "What made you pull my tail?"

The *Beagle* docked in the Galápagos Islands for five weeks, during which Darwin accumulated plants and animals, focusing on the many birds. He thought he was collecting blackbirds, wrens, and warblers, but when he got them back to London, an ornithologist told him that though the birds looked different they were all finches. Plus Darwin had stored birds in bags by type and hadn't separated much of his collection by island, which he later found was important. He'd assumed they were the same species he'd seen on mainland South America.

He did notice that the mockingbirds he'd taken on the second island seemed different from the ones on the first, so he started labeling them. When the vice governor of the islands told Darwin that he could distinguish the tortoises on one island from the tortoises on another, Darwin ignored him at first. Darwin did not imagine that these animals could have originated from a few animals blown across the Pacific and that they had diversified into different species on different islands within clear sight of one another. Darwin held, as did many scientists at that time, that these animals were all the same. Differences in color and form were indicative of different varieties, not separate species.

The definition of a species, according to Ernst Mayr, a German-born American biologist, is "groups of interbreeding natural populations reproductively isolated from other such groups." This definition didn't seem to fit the samples of wildlife Darwin had collected. These islands were in sight of one another. Surely separate species could not form on places so close. But they had indeed.

When Darwin returned to England, he gave all his bird skins and other trophies to the Zoological Society of London, and the ornithologist John Gould took a fresh look at them. At the next meeting of the society, Gould professed his excitement over Darwin's findings of a new group of "ground finches." The *Daily Herald* the next day reported on the meeting, noting the fourteen species of ground finches, "of which eleven were new forms none being previously known in this country." This finding heralded an important moment in the evolution of Darwin's *On the Origin of Species*, though it would be twenty-three more years before the book was published.

The fossils Darwin collected in South America were unique as well. Among them were a giant llama, a giant armadillo, and a rodent as big as a rhinoceros. Wherever one followed the trail of life, across the land or back through time, "species gradually become modified," wrote Darwin. He was beginning to realize how new species might evolve, but he had no idea at the time what a large role continental drift had played in the process.

On the voyage of HMS *Beagle*, Darwin brought *Principles of Geology* by Charles Lyell along for reading. Though his Cambridge professors had warned him to take the book with a grain of salt, he enthusiastically accepted Lyell's view of the earth changing restlessly beneath man. Darwin had witnessed this change in his journeys through South America. Still, both thought the movement of the continents was upward and downward, and that nothing moved laterally.

Darwin had no idea yet how important both the vertical and horizontal movement of the continents on the surface of the earth was to evolution.

GEOLOGY LED THE WAY

The mid-1800s were a time of upheaval in biological as well as geological thought. The British Empire was in full bloom and the most famous of the early geological surveys date from this era. The Industrial Revolution had arrived earlier with an insatiable hunger for iron, coal, oil, and other deposits, and thus geologists became the celebrities of the day. They earned their keep by uncovering industrial resources, and in accordance with the spirit of discovery that ruled then, these geologists weren't afraid to address more theoretical issues, like how these resources came to be.

Brothers William and Henry Blanford, members of the Royal School of Mines in London, were offered posts with India's newly hatched geological survey and were sent to investigate the Talcher Coalfield in the state of Orissa in that country. The Blanfords started digging and in 1856 found that below this enormous bed of coal was yet another formation of large boulders embedded in fine mudstone, and there was telltale evidence of a glacier. The boulders all had the markings of glacial scour—the abrasions, scratches, and polish of glacial ice against rock. Furthermore, some of the boulders had been moved large distances, another telltale sign of glacial action.

This showed that before Talcher had become one of India's largest coal deposits, formed by steaming tropical swamps, it had been part of an enormous ice field. The Blanfords returned to Calcutta and reported to their boss that ice sheets had once covered India. But this raised important questions in the geological community. How could glaciers form in the tropics? Had India once been much closer to the poles? Did continents move?

Further evidence for the shift of landmasses was uncovered in 1912 when Britain's Captain Robert Scott led a harrowing expedition to the South Pole, having to cope with blizzards and temperatures as low as minus 23 degrees. Though he and his men made it, they did so thirty-three days after a Norwegian team. Captain Roald Amundsen, its leader, left a Norwegian black marker flag and a note to the British at the pole. Losing the race for his country was enormously disconcerting for Scott, who wrote in his diary: "The POLE. Yes, but under very different circumstances from those expected. Great God! This is an awful place and terrible enough for us to have labored to it without the reward of priority." Scott and most of his men froze to death trying to get back, though they'd carried most of their finds almost the distance.

Scott's second in command, Edward Evans, survived, but upon returning to New Zealand wrote a letter criticizing their leader for not jettisoning all records and specimens of weight that the party had collected on their treacherous adventure. Scott and the team members Edward Wilson and Henry Bowers had died in a tent that was but 12.7 miles (20 kilometers) south of One Ton Depot, a spot on the Ross Ice Shelf where the party had cached food and supplies. Scott's body was found beside thirty-five pounds of coal and fossil rocks that the captain apparently considered more sacred than his own life. The samples included the first find this far south of *Glossopteris*, a seed fern that had become extinct over 200 million years ago. For such a tree to survive, a much warmer climate than the icy world Scott had found at the South Pole would have had to exist, scientists speculated. Or maybe the land that the South Pole stood upon had once been in the tropics?

Alfred Wegener, a German geophysicist, who first described the lateral movement of the earth's great landmasses in his 1915 book *The Origin of Continents and Oceans*, gathered evidence for this argument. Wegener noted that the continents of Africa and South America fit together quite nicely, and he found reports that fossils on the adjacent coastlines of both continents were similar. Scientists had previously suggested that land bridges once joined them, but Wegener countered this belief, saying that they had moved. He noted that India, Antarctica, and Australia looked like they could fit together, and proposed that they had all once been joined in a supercontinent that he called Pangaea. His book was the first place that name appeared. Wegener proposed that the world of today was but the dispersing remnants of this supercontinent, which 250 million years ago began to break apart.

Continental drift, or the slightly more evolved concept of plate tectonics, has been for scientists the driving engine behind evolution and the creation of new species for over a hundred years. In the days of the great Pangaea, all major landmasses had gathered together, and this merging of lands coalesced life—the outcome being fewer species. But as Pangaea began to separate, the isolation that followed proved the best breeder of species, creating a greater number of plants and animals.

ISLAND BIOGEOGRAPHY

There are, however, other ways to make new species. Alfred Russel Wallace, often credited with cofounding the theory of evolution, traveled through the Amazon and Southeast Asia in the mid-1800s. He studied hundreds of animals and tried to determine why they were found where they were. He thought it was significant that rivers and mountain ranges frequently marked the boundaries of species ranges. Many scientists believed that climate determined a species' range, but Wallace found similar climate regions with very different species and declared that geography had a lot more to do with it.

This theory of island biogeography, as promoted by Wallace and others, began as a way of explaining the species richness of actual ocean or lake islands, but grew to incorporate the species richness of landlocked islands as well. Scientists modified their definition of islands in the late twentieth century to include other isolated habitats such as mountains surrounded by deserts, lakes surrounded by dry land, and natural habitats surrounded by landscapes altered by man. Today, scientists have modified this concept further, using it to explain any ecosystem surrounded by divergent ecosystems. It could be an island surrounded by water, or a spring surrounded by desert, a mountain peak surrounded by lowlands, or grassland surrounded by human housing.

It is not a simple concept. What is considered an island for one organism may not be an island for another: some organisms located on mountaintops may also be found in valleys, being adaptable to both elevations. But others may be ecologically adapted only to the peaks and thus view the valleys as chasms that cannot be crossed. It may depend on whether the animal is a generalist, suitable to a wide range of environments, or a specialist, adapted to a much more specific niche. Isolated environments created in a mountain range can increase the variety of species in the range overall.

AN ISLAND IN THE ANDES

A typical example of a landlocked island is the mountainous region of Vilcabamba, the range in the Andes that I visited with Conservation International biologists. Deep river valleys surround the mountaintops, isolating them just like an ocean. The cloud forests here house many unique species, including some that have yet to be identified by scientists. Vilcabamba is a monument to natural diversity, as it showcases the broad range of possibilities to life.

On the day after Peruvian army helicopters had deposited our team into the dripping-wet cloud forests of Vilcabamba, I got up before dawn to survey the area's birds with Tom Schulenberg, an ornithologist with the Cornell Lab of Ornithology. We skirted the bog in the middle of the forest near our campsite looking for the feathered creatures, careful to avoid wet sinkholes in the moss that could swallow one's leg up to the thigh. Schulenberg aimed binoculars as well as microphones at the edge of the forest, claiming he could hear four times as many birds as he could see. Though the elevation here was too high for parrots and toucans, the ornithologist's Peruvian assistant, Lawrence López, captured a plush-capped finch, an Azara's spinetail, and a yellow-scarfed tanager that he pulled from his jacket pocket and proclaimed, "Look at that beauty," before he released it.

In the evening, I followed Mónica Romo, a biologist with the office of Conservation International in Lima, Peru, who set nets by the forest edge to capture bats that she estimates spread almost 50 percent of the seeds in the forest through their feces. The next day I followed Romo down trails freshly cut with a machete to lay mammal traps. Romo was knowledgeable about all these animals, but also bore a self-professed sweet tooth, and was envious when the camp's peanut butter was added to the bait used to capture these small creatures. "I hope they appreciate it," she said. Mindful of the presence of fer-de-lance, one of South America's most potent and aggressive vipers, I stowed my traps in the open, while Romo hid hers in every dark corner. They collected forty species of mammals, including a very large rodent that had never been described before.

A few days later, I accompanied Brad Boyle, a Canadian botanist who specializes in tropical plants. With his Peruvian counterparts, he laid a 165-foot (50-meter) line in the forest and started taking plant specimens on either side. He directed my attention to the orchids, bromeliads, mosses, and ferns perched on the limbs of the trees above, and declared that there were more species in that cluster of treetops than in most northern forests.

He told me how difficult it was to discover a new species. It was not like "Hallelujah! We've just found a new species!"; it was more like "We've just been through every similar-looking plant in the entire herbarium and can't find anything that looks like this." To declare something a new species requires a lot of work. Nevertheless he held up a tiny orchid a short while later and declared, "I'll bet a case of beer that's never been described before."

The rains carved the valleys that surround Vilcabamba, making it a de facto island, but not all lands are so easily separated. It took the massive forces of earth's molten core to break up Pangaea and to spread the continents wide. But now, through mass transportation, man has destroyed much of this hard-won separation. Similarly, as boats, trains, cars, and planes have spread across the land, so have the mammals, insects, reptiles, and crustaceans that have hidden in the bilges, trunks, and storage facilities of these vehicles. This has destroyed much of the isolation that created these many species, taking them to places where their presence can be ruinous to the resident animals.

In the United States the resultant spreading of some of these displaced species has been purposeful if not downright ridiculous. In 1890 an eccentric Shakespeare fanatic, Eugene Schieffelin, decided to introduce all the birds mentioned in Shakespeare's plays to the United States. Schieffelin released sixty starlings one fateful day in New York's Central Park, and from that introduction, the US now has 200 million.

These starlings, as well as sparrows and pigeons, make up the majority of the birds Americans see most days in urban environments. Yet none of these birds is native to the United States. "Invasives," as we call these exotic plants and animals, compete for food with true natives like eastern bluebirds and purple martins. Since local birds tend to migrate south for the winter, while invasives stay home, there is little nesting space available when the native birds return. And this is all because some dutiful citizen felt the New World would be more civilized if it were populated with the Bard's birds.

Invasive plants often flourish outside their native habitats because the insects, diseases, and animals that naturally check their growth at home are not present in their new digs.

The spread of an invasive species can also be caused by the inability of local animals to deal with new immigrants. The brown tree snake had such a free rein when it was introduced to Guam from the Solomon Islands after World War II. Scientists speculate that the snake probably snuck onto Guam inside the wheel well of an airplane, since Guam has an active air base. The brown tree snake spread across the island's jungles over the last sixty years and is responsible for the extinction or severe reduction of a number of native species that had no defense against the snake. Biologists recently attempted to control the snakes by air-dropping into the jungle dead mice laced with about 80 milligrams of acetaminophen—equal to a child's dose of Tylenol, all that's needed to kill an adult brown tree snake. The results are not yet conclusive.

Similarly, licensed and unlicensed animal traders over the last couple of decades brought Burmese pythons into Florida. When pet owners found that the snakes either took up too much space or tried to swallow the family dog, they let them go in local and national parks. Since 2002, more than 1,800 pythons have been removed from Everglades National Park and the surrounding areas in Florida. Now the US Fish and Wildlife Service reports that northern and southern African pythons, reticulated pythons, boa constrictors, and four species of anaconda have joined the Everglades pythons. Biologists believe that tens of thousands of these snakes now live in the park.

Invasive species can come from afar or they may grow up locally and penetrate spaces formally occupied by other species when the right conditions arise. Native woody shrubs and trees are invading semiarid grasslands in the US, South America, Africa, and Australia as a result of overgrazing, fire suppression, and climate change.

When animal grazing is controlled, grasses grow up naturally and provide kindling for natural or man-made fires, which stimulate more grasses but suppress the growth of woody shrubs. The control of woody shrubs, which inhibit grasses, is critical to pastoral communities on arid and semiarid lands, which make up 35 percent of the earth's people. They must balance the need for grasses to feed their cattle, sheep, and goats, with the need for fires to control woody vegetation.

To get a look at how shrubs outcompete grasses, I followed Rob Jackson, a professor of environmental earth system science at Stanford University, on a warm afternoon, down a tall ladder into the underground caverns of Powell's Cave, about 150 miles (240 kilometers) west of Austin, Texas. We entered a world of stalactites and stalagmites in the porous limestone bedrock of the Edwards Plateau in west-central Texas. I scurried after Jackson through a maze of caverns and tight crawl spaces, over slippery pathways, and into rooms filled with glistening multicolored limestone structures, all carved by nature. We arrived at a point about sixty feet (eighteen meters) below the ground, where an underground stream gushes from the rock.

Like the British geologists who went underground to look for evidence of glaciers, Jackson went deep to try to explain how native juniper trees on the Edwards Plateau have invaded grasslands and are taking over. He showed me several thick tree roots that appeared to burst from the limestone walls, reach down into the stream, and suck water out of it. He explained to me: "A single taproot can provide a third or more of the tree's water during a drought."

Junipers put out roots along the full depths of their root systems so that they can get water from deeper roots in times of drought, and from shallower roots in times of rain. This gives them an advantage over grasses, which can pull only from their shallow roots no matter what the weather.

Woody shrubs and trees like junipers have invaded arid and semi-arid grasslands and savannas in the US. Their presence limits the grass available for land managers, ranchers, and wildlife. Studies show that increased shrub and tree growth can rob from one-third to two-thirds of the stream water.

Juniper, mesquite, creosote, and Chinese tallow are problem plants in different parts of the US, particularly in the southern regions—the Great Plains, the Southwest, and the Gulf Coast. These plants existed here before, but overgrazing, fire suppression, and climate change have allowed plant populations to explode. Their increased presence has led to thickets that don't allow enough light or room for other plants or grasses. "Thicketization" was what Steve Archer, a professor at the School of Natural Resources at the University of Arizona, Tucson, called it when I met him at an Ecological Society of America conference in Austin, Texas. Archer studies the ecology, management, and restoration of rangelands, which are any extensive area of land occupied by native vegetation that is grazed by domestic or wild animals that eat plants.

When huge herds of cattle were introduced to the western United States in the late 1800s, they devastated the grasses. This reduced the fuel for grass fires, and in their absence woody plants got better established. In earlier times, Indians regularly burned the grassy meadows here to clear brush and trees and open them up for hunting. Today fire suppression is one of the problems promoting the juniper invasion. Without grass fires, woody plants spread unabated.

But it's more than a recent problem. It goes all the way back fifteen thousand years ago to Ice Age hunters, who wiped out the large animals that once ate the woody plants in the grasslands of North America. In East Africa, they still have elephants that control woody plants, one of their natural foods. But the US no longer has wildlife populations that can do the job, so woody vegetation grows uncontrolled.

Woody plants in the US can be bad for ranchers, farmers, and wildlife. Black-capped vireos and golden-cheeked warblers, both considered endangered in Texas, are two species that need a mixed landscape of forest *and* open grassland to thrive. Woody vegetation suppresses the grasses and the open space that comes with them.

An example of the woody plant problem can be found at the Tallgrass Prairie Preserve near Pawhuska, Oklahoma, the largest protected area of tallgrass prairie in North America. Tallgrass prairies once spread throughout the Midwest, supporting enormous buffalo herds, and though there are still buffalo here, their numbers are small. Private ranches surround the preserve, and land managers at these ranches find that if they don't burn grassland areas every year, woody vegetation forms canopies, which makes them immune to future fires.

Woody vegetation is also invading normally bald mountaintop environments in New Mexico as well. Here, bighorn sheep typically gather because they can see mountain lions approaching and escape. But as woody vegetation moves into these formerly bald areas, it allows mountain lions some cover from which to attack, wreaking havoc on bighorn populations.

"Climate change, increasing atmospheric carbon dioxide, and on-the-ground changes like fire suppression and cattle grazing should speed the global transition to woody species," says Jackson. "It's not just a problem in Texas, but in South America, Africa, and Asia as well."

* * * *

Climate change, its causes and effects, is another issue Jackson is going underground to work out. He is currently looking at problems with natural gas, which was once thought to be an ideal solution for some of our greenhouse problems, since it burns cleaner. It's certainly a cleaner fuel than coal or oil, but Jackson is concerned with leaks that occur in transport. Belowground fracking can result in leakage into groundwater on the extraction side and old pipes can sprout leaks into urban soils on the delivery side.

Jackson and the Boston University professor Nathan Phillips found natural gas (methane) escaping from more than 3,300 leaks in Boston's underground pipelines, where there is a record of natural gas blowing up homes, regularly sending manhole covers into the sky, and killing trees.

Still, despite all the importance given to greenhouse gases, Jackson thinks the spread of invasive species across the globe is more permanent and perhaps a more serious threat to our environment. We can reverse climate change in one thousand to ten thousand years, but the plague of invasives and the mixing of species worldwide is not one we are likely to recover from.

4

EVOLVING OUR WAY TOWARD ANOTHER SPECIES

THE AUTHOR OF MOST of our conflicts with nature is our own species. But it wasn't always that way. For an idea of how we once coexisted with the land and its animals, we visited Olduvai Gorge, part of the African Rift Valley near the Tanzanian border with Kenya. The gorge is a place where many have come to understand how man developed intelligence, learned to talk, and eventually spread over the world, his numbers exploding in recent years. The idea of a future species of man seems fanciful to many now, but in or near Olduvai there is evidence of three other species of hominids distinct from *Homo sapiens*: *Paranthropus boisei*, *Homo habilis*, and *Homo erectus*. Their existence shows not only how we got here but how a world of one hominid may not be that natural after all.

A giant plume of magma pushes the land upward, lifting Olduvai Gorge to an elevation of four thousand feet. Even though it is close to the equator, the weather here is mild. In late June, it ranges from daytime highs in the seventies to nighttime lows in the fifties and sixties, which is typical even into the dry season.

The morning after my arrival the sun broke over low shrubs, thorny trees, and savanna grasslands that cover the dry landscape.

Most of the vegetation here evolved with large animals and early man, and brandished nasty spikes or spines to discourage plant eaters. I was with a group of anthropologists, geologists, and paleontologists at a field camp hosted by the University of California, Berkeley. I emerged from my Quonset hut in the former camp of the notable anthropologists Louis and Mary Leakey, among the many scientists who have studied here.

There were field camps from various international institutions here. The Leakey family made Olduvai Gorge famous beginning in the 1930s with unique findings of various species of hominids. Much of the attention devoted to this place was brought by those hoping to find similar fossils and similar fame.

At our camp, scientists from all over the globe rose with me to welcome the sun and start the workday. Our field site comprised a number of corrugated metal buildings and tents that were the base camp for about twenty anthropologists and paleontologists and their Maasai tribal assistants. We consumed a hearty breakfast of millet, porridge, eggs, fresh-baked bread, assorted fruits, and lots of coffee before we got into a half dozen safari vehicles and headed out for the day.

Leslea Hlusko, a professor of paleontology at the University of California, Berkeley, drove our safari vehicle in a caravan with several others past the field site of the Spanish scientists, who waved at our car. The Spaniards acted friendly, but Hlusko assured me there was a competitive fervor among the various international groups at Olduvai. Everyone wants to make a difference, but with the history of historic finds credited to this place, it is hard to find room in the spotlight.

Hlusko is codirector of the Olduvai Vertebrate Paleontology Project, which is trying to develop an online database of fossils so that scientists can readily access past projects and know where fossils are stored. "We want to make the data from the projects available to everyone, and also let them know where the fossils are located, whether they be in a museum in London or in someone's basement in Florida," said Hlusko.

Hlusko was also trying to identify the genetics in the fossils, utilizing a unique reverse analysis. Part of her past work had been with captive groups of baboons in the United States, studying their teeth and then identifying what genes were responsible for their placement, size, enamel, and dental surfaces. Hlusko hoped to study the fossils here and then to determine the genes behind the baboon teeth, including complementary characteristics in other parts of the body that those genes might have turned on. "We know a lot of hominids and early primates just by their teeth or some part of the jaw—particularly as you go further back in time," Hlusko pointed out.

The project was interested in the effects animals in this ecosystem had on the hominids who were once here with them. At one time the Olduvai Gorge area was a place where animals interacted with man in a balanced community. We saw that some of this balance still exists today, since this area is surrounded by national parks.

We continued traveling along the ridge of the gorge until we came to a plateau where the team parked its vehicles and the men and women aboard prepared to go to work. As you looked out over the gorge, you could see the layers of earth in the sides of the canyon. We were fortunate enough to have these well-defined layers, which Hlusko said made it possible to determine the era of a fossil by the stratigraphy of the soil. The group of scientists and Maasai helpers spread out over the sloping side of one section of the canyon. I learned to avoid those climbing precipices after I followed one group up a pinnacle and had trouble getting down.

That morning, we found the lower jaw of an ancient mastodon, and Hlusko spent more than an hour extracting it from the ground, carefully packing it into a plaster mold to take back to camp. She explained that she normally avoided hippo and elephant bones because they don't appear to evolve as much as man or some of the other carnivores. But this elephant jaw was so intact that she just couldn't resist.

* * * *

Later that week the project's codirector, Jackson Njau, took me to his study site in Serengeti National Park along the Grumeti River. Like Hlusko, Njau was interested in how animals related to early man, but he had chosen to focus specifically on crocodiles and their possible effect on hominid intelligence. Trees and brush lined the river, though savanna grasslands dominated the greater landscape. Njau was born in Tanzania and got his BA there at the University of Dar Es Salaam and his PhD at Rutgers University before taking a position at Indiana University. He and Hlusko had worked together at other sites in northern Africa.

We arrived on a cloudy day at the Grumeti River, where more than twenty hippos weighing 3,000 to 10,000 pounds (1,600 to 4,500 kilograms) glistened in the sun as they jostled with each other for a place in the water. Along the shores lurked four or five crocodiles, their rough, bumpy skin and long, toothy jaws blending eerily into the landscape. Though the hippos were not to be ignored, it was the crocodiles that drew the most attention from Njau.

According to Njau, crocodiles are the most dangerous predators in Africa, causing far more deaths than lions or leopards. Crocodiles have killed more than five hundred people in Tanzania alone since 1985. Njau explained the inevitability of these occurrences: "Victims know where they are, how to avoid them, yet they still keep getting caught and killed." Njau warned me that for every crocodile you see above the water, there can be several others below waiting. I stayed far back from the river.

The reason for this caution was that crocodiles frequented the water where men or animals came to drink or bathe, and these secretive reptiles were very, very patient. At just the right time, when its prey, man or beast, came forward, convinced there was no danger, the crocodile would lunge out of the water with extraordinary speed and grab its victim. The crocodile locks its jaws on the head, shoulder, arms, or front legs of its prey, then drags its spoils back into the pool to be held down to drown.

During his thesis study, Njau came to the Grumeti River in the early summer to observe how the crocodiles overtook other animals. He visited a few months later in the dry season when the river had vanished and the crocodiles and hippos had moved on. He studied the bones left in the middle of the pool where the river had dried out and compared the tooth marks left by crocodiles to those of other carnivores. He wanted to know what marks the different predators made so he could study fossils and better know what was happening back then.

Crocodiles had rows of as many as sixty-six teeth along powerful jaws that were ideal for gripping prey. The crocodile would often grab a victim and beat it against a rock, or it would sometimes go into a death spiral and roll over and over, or two crocodiles would grab the victim and roll over in opposite directions. The crocodile tried to disarticulate a substantial chunk of meat and then swallow it whole, allowing the reptile's stomach acids to do most of the digesting.

Crocodiles left puncture wounds where their jaws locked on a victim. But they weren't able to move their jaws from side to side. This meant they actually made fewer marks on the bones of prey than other predators, although in some instances they left dense concentrations of bites on bones they were unable to swallow. A crocodile would tend to rip a large piece of meat and bone from a victim, swallow it whole, and then leave the rest. Lions, leopards, and even hyenas would gnaw on the ends of the bones to get the meat off, and even break the bone to get at the marrow. Thus most of the bones of crocodile victims would show fewer tooth marks than the bones of lion or leopard victims, and lacked gnawing damage on the ends. Crocodiles would take only right-size chunks of meat—not too big to catch in its throat, but not too small not to warrant the effort of the hunt. Bones that didn't fit either of these categories were left in the water to settle to the bottom of the pond, along with leftover crocodile teeth.

Back at Olduvai on a day when the sun was beating hard, we revisited the high ridge of the gorge, parked our cars near a cliff, and hiked down the cliff face to a wide bend in the dry creek, a site surveyed by the Leakeys in the 1960s and more recently by the current crew.

Codirector Njau used his work in the Serengeti on living animals and their prey to understand how crocodiles might have affected early man in Olduvai. Njau told me that the entire gorge area had been through a series of drastic climate changes since two million years ago, when *Homo habilis* first occupied it. He said it was much more humid then, and that there was a lake not too far downstream from where we stood into which the river flowed. Today, the area was very dry with the exception of a few months of the rainy season when the Olduvai River floods briefly.

Homo habilis took meat here, but not as a hunter. "Most people think, 'Oh, here comes man, he must have been a hunter,'" said Njau, "but no, *Homo habilis* was only about three or four feet tall, and weighed less than one hundred pounds. There was no way he could have brought down wildebeest- or gazelle-sized animals here. We think he lived here as a scavenger, feeding off the kills of lions and leopards."

In order to survive like this, he had to be well aware of the terrain, and he had to forage in groups. One man alone was too much of a target for local predators. The Leakey family originally hypothesized that early man may have lived along the waters running through Olduvai, but Njau thought that the evidence for crocodiles, hyenas, leopards, and lions as well as hippos, elephants, and other wetland animals was too great and the result too dangerous for men to have stayed here on any type of permanent basis. He considered this space more as a puzzle that *H. habilis* had to learn to navigate, and that an ability to plan, to hunt cooperatively, and to anticipate predator movements would have provided selective pressure for early human intelligence. If you didn't figure out the puzzle, you died and didn't pass on your genes—the basic element of evolution.

Fossil evidence for tool use at Olduvai showed that *H. habilis* had instruments for cutting flesh away from bones, but not stone spear- or arrowheads that could kill fully grown animals or drive off lethal predators. They may have had wooden spears, but they were scavengers, not hunters. Being smart here improved their chances of survival but also provided access to meat, which had the extra calories necessary to evolve bigger brains.

Studies of the lineage of man provide not only a look at the past but also insight into the future. It was only in 1856 that scientists found the fossilized bones of the first extinct human in the Neander Valley, Germany. Darwin mentioned nothing in *On the Origin of Species* about the evolution of man, though the book came out in 1859, three years after the first Neanderthal fossil was discovered. It was not until 1864 that these fossils were recognized as a separate species, *Homo neanderthalensis*.

Darwin later proposed that our early ancestors diverged from Old World monkeys in the early Miocene epoch, about 20 million years ago. In 1927, Dr. H. L. Gordon, a retired government medical officer, found in limestone deposits in western Kenya a specimen of one of the first primates to diverge from apes. Gordon named it *Proconsul africanus*, after a chimpanzee named Consul that performed in the *Folies Bergère* in Paris in the early 1900s. The *Folies* chimp wore a tux, played the piano, and smoked a cigar, before taking off his trousers, standing on his head, and somersaulting into bed. Mary Leakey found one of the most complete skulls of *Proconsul* in 1948 on Lake Victoria.

The discoveries of early primates leading up to man didn't end with *Proconsul*. Africa had many more tales to tell. *Australopithecus afarensis* (better known as Lucy, which lived from 3.85 million to 2.95 million years ago) was discovered in 1974 by American paleontologist Donald Johanson and grad student Tom Gray. They found Lucy in Hadar, Ethiopia, and celebrated under a star-filled sky playing the Beatles song "Lucy in the Sky with Diamonds." Lucy walked upright with a humanlike pelvis but had a small brain and primitive teeth. She had a powerful jaw, probably used more for stripping plant material than eating meat.

Then meat eaters started to appear. *Homo habilis*, the scavenger, was the earliest of the genus *Homo* to which we belong. *Homo habilis* (2.33 million to 1.6 million years ago), the "handyman," was so named from early tools found by the Leakey family at Olduvai. The males stood about 4 feet 3 inches (1.3 meters); the female stood about 3 feet 3 inches (1 meter). *Homo habilis* was short but had a significantly larger brain than Lucy's family. Its carnivore diet provided the calories that enabled that growth.

Homo erectus (1.8 million to 140,000 years ago) is thought to have evolved from *Homo habilis* in Africa. Its fossils follow *H. habilis* in the stratigraphy. The first findings were made at Trinil, Java, Indonesia. German evolutionist Ernst Haeckel had predicted that the origins of man would most likely be found in Southeast Asia. It was largely through the work of the Leakey family that scientists began to recognize Africa as the more likely birthplace of both *H. habilis* and *H. erectus*.

Still, if *Homo erectus* did evolve from *Homo habilis*, he did so in one amazing growth spurt. If you look at the life-size replicas of *Homo habilis* at the Smithsonian's Hall of Human Origins, *habilis* is a little guy with not much heft. *Erectus* appears as if he's ready for some early hominid basketball. Standing at 6 feet 1 inch (1.8 meters), he's looking down, I must confess, at me. Fossil remains of *Homo erectus* are found predominantly from 1.8 million to 140,000 years ago. He was the first of the hominids to migrate to the Far East and to Europe.

Homo sapiens evolved in Africa about 200,000 years ago from *Homo erectus* via one or two intermediate species. *Homo neanderthalensis* evolved via its own intermediaries in Europe. About 120,000 years ago, early *Homo sapiens* migrated out of Africa to the southeastern Mediterranean coast, infringing for a while on Neanderthal territory before a cold phase ensued and *Homo sapiens* pulled back into Africa. The next time they appeared they were better equipped and more numerous.

* * * *

Back at Olduvai, Tomos Proffitt, a graduate student at University College London's Institute of Archaeology, sat on a bench outside the field house trying to simulate what it must have been like for early man to make stone tools. He held a single round rock, or "hammerstone," in his right hand and a larger piece of rock cradled in his left hand resting on his knee. He took careful aim with his hammerstone before bringing it down at an angle on the larger rock with sufficient force to break off chips and extend the sharp edge around the larger stone. A pile of rounded rock chips surrounded his feet.

Proffitt's hammerstone was made of quartzite and the large stone, the eventual hand ax, was made of phonolite, a fine-grained lava rock. *Homo erectus*, the tall guy, used similar tools to butcher meat and possibly to sharpen wooden spears.

Simple flakes, small pieces of sharp rocks used for cutting, are thought to be a part of the *Homo habilis* tool kit. Bifacial tools, like the hand axes Proffitt spent hours each day crafting, are part of the *Homo erectus* tool kit. It was in the 1970s that the first tools were recognized in Olduvai Gorge, dating back to 2.5 million years ago. Their creation shows that early man had the mental capacity, the dexterity, and the fine motor skills to craft tools as well as use them.

The tool technology of early man plays a determining role in why Neanderthals ended up on the losing end of their mortal competition with *Homo sapiens*. Neanderthals, our closest relatives, dominated Eurasia for the better part of 500,000 years, spreading over Europe, Britain, Greece, Russia, and Mongolia. Despite the broad reach, their population is thought to have been from 10,000 to 100,000 total individuals at its apex. Examinations of Neanderthal bones reveal that adult males had greater strength in their right arm as opposed to their left, indicating that they carried heavy, hand-held spears, which they probably used for thrusting rather than throwing. Males had solidly built bodies, standing only about 5 feet 5 inches (1.7 meters) but weighing around 185 pounds (84 kilograms). They needed 5,000 calories a day to do their job, an amount approximately equal to what a typical cyclist needs to compete for a day in the Tour de France.

Neanderthals hunted in forested areas where they could ambush prey at close range using thrusting spears to bring them down. They relied almost completely on large and medium-size mammals like horses, deer, bison, and wild cattle for food. It was a hard living. Remains of Neanderthal bodies resemble rodeo contestants in that they bear multiple scars and fractures. Neanderthals adapted to live in warmer forested climates, but toward the end of their reign, Europe got colder and ice covered Scandinavian mountains and northern Britain, enclosing them in barren, glacial landscapes. Neanderthals moved into the southernmost forests surrounding the Mediterranean to escape the cold and the spreading open terrain.

Meanwhile, *Homo sapiens* moved up into Europe during a brief interval in the larger cold phase between 58,000 and 28,000 years ago, and adapted well. They were lighter than Neanderthals, needed fewer calories to survive, and were more omnivorous. And they weren't against an occasional meal of fish or even a vegetable or two. They hunted with lighter stone-tipped spears that could be thrown at a distance. They also used spear throwers (atlatls), which consisted of a shaft with a cup at the end into which the spear fit, giving the hurtler and his spear additional leverage, distance, and velocity. Using a thrower, a human could toss a spear up to 325 feet (100 meters), though it was most effective and mortal at about half that distance.

The end came for Neanderthals during the period when *Homo sapiens* were undergoing their grandest growth—an explosion of culture, symbolic communication, and art. While the communities of modern man grew, Neanderthal communities remained small, and the range of their influence was smaller, too. Neanderthals were lucky to collect stones sixty-two miles (one hundred kilometers) away, while *Homo sapiens* collected stones up to three hundred ten miles (five hundred kilometers) away.

"*Homo Sapiens* had the ability to develop trade at a much greater distance than Neanderthals," says Rick Potts, director of the Smithsonian's Human Origins Program. "Our species can get something five hundred kilometers away, or develop an alliance with someone five hundred kilometers away. In the end, that can really buffer bad times."

Neanderthals gradually disappeared from Britain, Greece, the Middle East, Russia, and Mongolia. Their last stand may have been in the caves beneath the Rock of Gibraltar. Did the Neanderthals go peacefully, or were they pushed over the evolutionary cliff? Scientists believe that early hunter-gatherer societies were more aggressive than previously judged.

One of the biggest advantages *Homo sapiens* may have had over other hominids was language. Language gave moderns the ability to pass on the lessons of the past. Communication and the ability to remember and utilize a broader range of information allowed for innovation. "And *Homo sapiens* had the ability to accumulate innovation," says Rick Potts. No other species had demonstrated this before.

The ability to understand language may have actually preceded the ability to talk. Beneficial mutations in the genetics of understanding may have come before mutations in the ability to mouth words. Thus scientists have had luck getting primates, chimpanzees, orangutans, and bonobos to understand language but less luck getting them to talk.

Robert Shumaker, vice president of conservation and life sciences at the Indianapolis Zoo, told me that there have been several studies aimed at getting monkeys and other primates to speak English. Vicky, a chimpanzee raised by scientists in the early 1900s, was trained to vocalize breathy imitations of "Mama," "Papa," "cup," and "up," but the efforts were laborious and the lessons soon forgotten. Great apes simply don't have the morphology to form English words. However, Bonnie, an orangutan that Shumaker worked with at the National Zoo in Washington, DC, demonstrated a series of whistles that she used to let caretakers know her needs. "And she did this without training, essentially creating her own vocabulary and syntax to go with it," says Shumaker.

Experiments at the think tank at the Smithsonian National Zoological Park in Washington, DC, display this innate ability. Chikako Suda-King, on a postdoc fellowship with the David Bohnett Foundation at the Smithsonian, took me behind the scenes one autumn day to meet the famous Brainy Bonnie, the orangutan that Shumaker had worked with. When Suda-King walked into the cage area, Bonnie started to act like an excited kid, but soon grew serious when she saw Suda-King roll a computer up in front of her cage. She promptly settled down in front of her computer screen.

Suda-King was trying to determine if Bonnie had the ability to make decisions based on the perception of her own knowledge of a given subject. In other words, was she capable of asking herself: Do I know enough to take this test? This is a level of self-awareness that was previously thought unique to humans. On this day Suda-King presented five pictures, all of the same thing, which the orangutan individually tapped to move forward in the game. The next screen gave Bonnie a choice of two pictures presenting options that she had learned to translate as follows: Do you want to go to a second test of your recall of those photos and get three grapes, or do you want to opt out of the test and take only one grape? If the animal chose to take the test and missed, she got nothing. During the test phase, she was shown several photos; only one was similar to those in the study phase.

Bonnie picked the test over and over, and consistently matched the correct photos. Suda-King, who holds a PhD in animal psychology, had to slip Bonnie her three-grape rewards through a slot between them, and today Suda-King was having trouble keeping up with the orangutan. "We're going to have to make this test harder for her," she joked, though she admitted it had actually taken Bonnie a couple of years to figure the test out. Still, it proved self-awareness in a primate.

Sue Savage-Rumbaugh, a research scientist at the Great Ape Trust in Des Moines, Iowa, claims her bonobos, relatives of chimpanzees, are able to communicate with caregivers through human sign language and a system of lexicons. Savage-Rumbaugh says her bonobos are allowed to watch soap operas, which they select and turn on themselves, and they have shown a preference for sequential stories.

Lisa Heimbauer, a doctoral candidate at Georgia State University, taught a chimpanzee called Panzee to understand English by treating the chimp like a human from shortly after it was born. Panzee can currently understand 130 human words even when those words are offered in computer-distorted speech that was thought to disguise those words from anyone other than humans. Heimbauer believes that primates developed understanding before speaking. "The cognitive abilities to perceive speech had to be there when production evolved," she told me in a telephone interview.

Though both Neanderthals and *Homo erectus* are thought to have had some form of basic speech, *Homo sapiens* were better at acquiring and advancing it. This gave them a dramatic advantage over other hominids because they could engage in trade while learning to navigate the wildly fluctuating climate of the ice ages. In *The Third Chimpanzee: The Evolution and Future of the Human Animal*, Jared Diamond describes this as the "Great Leap Forward," or the dawn of culture. Speech involves a series of mental images or symbols that represent words and thoughts. Though there is no direct evidence of speech in Neanderthals, there is evidence of common tool use that must have required some sort of communication between the different hominids, and possibly interbreeding. The latter may be responsible for shared genes in both species, particularly the FOXP2 speech gene, which anthropologists say *Homo sapiens* may have picked up from Neanderthals.

Speech may have given modern man the same advantage that large lungs gave *Lystrosaurus*, the bulldog with tusks, during the Permian extinction. Like man, *Lystrosaurus* grabbed this advantage and populated much of the world, as did the newly evolved dinosaurs after the Triassic extinction: they seized the advantage from the then-dominant crocodile-like predators, and soon ruled the world.

Today, modern man is the planet's most successful creature, occupying virtually every environment on earth except the deep ocean and the polar ice caps. But our population growth of *Homo sapiens* has reached a zenith in the last fifty or sixty years, and we are now at the point where our celebrated progress has become our greatest nightmare.

The population boom of Los Angeles, California, shows how growth can rapidly accelerate with little notice by the residents but with great consequences for the environment. The town was established in 1781 when the Spanish governor at the time convinced 44 people to come up from Mexico to investigate the possibilities of this new and untrammeled land. By 1800, the 44 people who had settled there grew to 315 people. By 1850, after Mexico had ceded California to the United States, there were 1,610. By 1900, there were 102,479. Then they found oil in some of the beach towns. In the 1910s and 1920s the film industry moved from New York out to the West Coast for better weather. With the breakout of World War II, they started building planes. By 1950, in an area of about 502 square miles—smaller than London or Tokyo but bigger than New York—there were 1,970,358 people living.

I grew up in Los Angeles's Westside when it was mostly single-family dwellings. Traffic on the street was light, and there were few freeways. Then the government started building freeways and moving my friends out, buying up their property for public use under the government right of "eminent domain." Over time, the single-family dwellings turned to multiple dwellings. The former apartments turned to apartment towers. My family used to drive east toward the mountains or the desert and we would gaze at orange groves all along the way. Now there are homes, apartments, car lots, and mega-malls.

Today the population is close to 3.9 million. And most of this growth, both the human population and the infrastructure, developed in the last one hundred years.

A similar tale of growth is true for New York City. When the surveyor John Randel Jr. submitted his intricate grid of the streets of Manhattan in 1811 that would eventually develop into Greenwich Village, SoHo, Times Square, and all its famous communities, this central isle surrounded by rivers was but a New York City borough of eighteenth-century villages. "The island was hilly and stony, woven with creeks, soft in places with beaches, marshes, and wetlands," writes Marguerite Holloway, author of *The Measure of Manhattan*, the story of Randel's grand achievement. In 1800 the city's population was 60,000. Today the Census Bureau shows New York is the most populous metropolitan area in the US, with an estimated 8.4 million residents.

In those two hundred years, London has grown from about 960,000 to 2.8 million. In one hundred years, Tokyo has grown from 3.7 million to 13.2 million. Istanbul grew from 3.7 million to 13 million. When you include the immediate suburbs around these cities, estimates can more than double. As the population of the world has risen, so has the population of its great cities.

If we were to make a chart of the world's population growth, it would look a lot like the Keeling Curve, which shows the growth of CO_2 in the atmosphere in the last thousand years. It's often referred to as a hockey stick, because CO_2 emissions remained steady for most of this period until 1850, when the Industrial Revolution swung into gear worldwide and amounts grew from 280 ppm (parts per million) to 396 ppm today. The world population in AD 1 was about 200 million. It increased slowly over the first millennium, but started accelerating in the second millennium, particularly toward the end. By around 1800 it was at 1 billion. By 1930 it was 2 billion; 1987, 5 billion. By 2011, 7 billion. By 2024, it is predicted to reach 8 billion, and by 2045, 9 billion. If the world continues to grow as it has in the last fifty years its population could reach 27 billion by the year 2100, which is unsustainable. Among a ton of other issues, there simply wouldn't be enough food to feed that many people.

Population experts are relying on a radical slowdown of this growth brought on by national controls, but also by women holding off childbirth to take advantage of increasing opportunities in industry and education. Many scientists project that population growth could start to level off during this century, and by 2100 we may be only ten billion. Still, that figure represents three billion more people on earth than there are today.

The largest population growth has occurred in Asia and Africa. From 1960 to 2011, India gained 782 million people, the single largest contribution to the planet's population in the world. By 2030, India's population is expected to top China's. In India, women have an average of 2.5 children each. But by 2030 the rate is expected to fall to 2.1 children, close to the replacement rate of 2 per couple. The problem is that over the last fifty to sixty years we've come through the largest population boom in the history of man, and this has momentum. United Nations demographers used to think it would peak in 2075. Now they say it will continue to grow into the next century.

There are cultural barriers to population control. Fertility is still advanced by tradition, religion, the lower status of women, and limited access to contraception. In the US, support for family planning has actually dwindled. Population growth today gets less attention than it did in the 1960s, when there were only half as many people alive.

India's population growth has slowed among the urban middle class but remains high among the rural poor. Producing a male heir is still a family tradition in the Hindu culture. Sons provide for their parents in old age and perform last rites, a duty seen as necessary for access to heaven. Having ten or fifteen grandsons is still considered a healthy sign by some Indian grandparents.

China's one-child policy has slowed growth there, but the momentum created by hundreds of millions of young Chinese still in their reproductive years continues to be a force in the other direction. The country's current population is 1.3 billion. Today young couples are given housing subsidies as well as retirement benefits if they hold to one child. But even if they don't have more children, they are still consuming larger portions of the country's resources in terms of food, energy, and goods as their economy grows.

Population growth in Southeast Asia and the Middle East fuels strife. In rapidly developing countries, young men can't find employment. However, some can make a living ambushing military vehicles or foreign supply trucks and dividing up the food, blankets, and spoils. Population experts say that 80 percent of the world's strife since the 1970s has been driven by the explosion of youth.

Without employment, young men can't save money for a dowry, without which many in the Middle East and Southeast Asia can't get married. And their culture places heavy penalties, even death, on sex outside of marriage. An enormous youth population denied employment, money, and sex is a formula for disaster.

Two billion more people may be added to the world's population by 2050, and most of them will come from the poorest countries in Asia, Africa, and Latin America. Security assessments by the US National Intelligence Council warn that climate change could harm food, water, and natural resource supplies, which in turn could lead to global conflicts.

Some say population growth may level off in the next hundred years, but this wouldn't include human consumption of natural resources. Third World nations that are becoming industrialized want some of the benefits afforded to those that arrived earlier—like cars, electronics, and meat. Modern estimates for how many humans the world could support range from a high of 33 billion for people fed on minimum rations only, down to 2 billion if they all lived like middle-class Americans, a style of life many pursue because they watch it on television.

The Population Bomb was a book published in 1968 by Paul R. Ehrlich, a Stanford University professor, that warned of mass starvation as the result of overpopulation. The book was criticized for being alarmist. The cover of the book had this statement. "While you are reading these words four people will have died from starvation. Most of them children." Ehrlich and his wife, Anne H. Ehrlich, recently revisited the population issue in the *Electronic Journal of Sustainable Development* and concluded that the message of *The Population Bomb* is even more important today than it was forty-five years ago. "Perhaps the most serious flaw of *The Bomb* was that it was much too optimistic about the future," they wrote.

Their view is that humanity has reached a dangerous turning point in its domination of the planet. We are increasing our numbers and our appetites for our natural resources at the same time. This behavior simply cannot continue.

One of our looming problems is our soil, a vital resource needed to feed our exploding populations. But is the earth's soil ready for the job?

Part II

WARNING: DANGER AHEAD

5

WARNING SIGN I: THE SOIL

THE RISE OF AGRICULTURE about ten thousand years ago contributed to the historical population growth of man, but along the way we've been destroying the very soil we now desperately need to feed our growing population in the future. To get some perspective on this problem, I visited a number of spots around the world, including Rothamsted Research, the longest-running agricultural research station in the world. The institute, founded in 1843, is in the town of Harpenden in southeastern England, about thirty miles north of London, once dominated by stands of lindens mixed with oak and hazel. This land was transformed into grasslands around 4500 BC when immigrants crossed the English Channel and brought in domestic crops and animals.

I stepped off the train as clouds were receding in the sky. The village was flush with bright-green grasses, roughened sidewalks, and colorful ornamental flowers stippled with the early morning rain. Small agricultural fields surround the hedge-bordered town. I walked several blocks to the facility where I hoped to learn more about what the history of agriculture had done to the soils on our planet and what we might expect from them in the future.

Rothamsted Manor, solidly built with bricks and old timbers, lies in the middle of three hundred acres of rolling verdant agricultural lands a few blocks from the train station. The manor was first mentioned in historical documents dating back to the early 1200s. Since then it has grown, rooms have been added, and the title has changed hands at least five times.

John Bennet Lawes, entrepreneur and agricultural scientist, assumed responsibility for the management of Rothamsted Estate in 1834 after leaving Brasenose College, Oxford. He started a number of agricultural experiments indoors and in the field. Justus von Liebig, a German chemist, taught him how to take bone and boil, grind, and treat it with acid to make fertilizer. Lawes was soon selling "superphosphate" made of powdered phosphate rock treated with sulfuric acid to all the locals. Superphosphate was a super success.

In 1843 he started a fertilizer factory in London, got married, and appointed Joseph Henry Gilbert to manage the field experiments—the official start of the Rothamsted Experimental Station (later Rothamsted Research).

To determine the best methods, Lawes and Gilbert planted two fields of wheat and turnips, divided these into twenty-four strips, and then applied different fertilizers and chemicals to each, refining the ingredients over time and adjusting them for different crops until they got optimum growth. They also took notice of the different effects on crop yields from inorganic and organic fertilizers and how they affected the biodiversity of the plants and animals around them.

Inorganic fertilizers came from mining or mechanical processes. Organic fertilizers came from animals and plants. Lawes and Gilbert determined that all plants increased yields with the addition of nitrogen and phosphorus, whether inorganic or organic. Trace minerals increased yields on some plants but not others. They added fish meal and animal manure from a variety of animals, which were fed different diets. In 1889 Lawes established a trust from the sale of his fertilizer business, and so experiments continued after his death in

1900. Rothamsted researchers began to test the pH level of the soil to determine its acidity or alkalinity, and then added chalk to vary those results and test the difference it made.

In general, fertilizers accelerated crop growth, but researchers also noticed that where inorganic fertilizers were added, waist-high crops grew on agricultural lands, but in nearby fields, species numbers declined. Up to fifty species of grasses, legumes, weeds, and herbs grew on fields away from treated lots, but as few as three species grew on lots adjacent to fertilized plots. Inorganic fertilizer improved crop yield, but it dramatically decreased biodiversity. The arrival of agriculture initiated the long decline of plant species numbers and is part of the biodiversity crisis we are now experiencing.

The spread of agriculture compounded the effects of man on nature and its suite of plants and animals. Most farmers went along with the reduction of biodiversity, since a decrease of plant species led to fewer weeds. Still, modern agriculture has brought down the number of plant species on earth much like the volcanic eruptions that triggered the Permian extinction and the asteroid that ended the Cretaceous. Fewer species leads to the spread of disease by reducing the number of hosts that carry disease, some of which are better at spreading it than others.

Lawes had trepidation about the situation he had helped create. Though he had been one of the prime movers of inorganic fertilizers, he advised anyone planting vegetables or garden greens to find a location near a farm with a "large supply of yard manure at a cheap rate." Organic fertilizers, particularly manures, were a better choice even to the inventor of many of our inorganic choices.

In the late 1800s, Europe was struggling to feed its burgeoning population as farmers desperately sought manures for their grains and vegetables. South Pacific islands were stripped of their guano; stables were ravaged for the smallest of droppings; and human refuse, delicately referred to as "night soil," was tested as well. According to Liebig, even the horse and human bones (good sources of phosphorus) from the Battle of Waterloo were ground up and applied to crops.

Inorganic fertilizers were thought to be the only logical choice by the dawn of the twentieth century, and Queen Victoria knighted Lawes and Gilbert for their agricultural innovation and the benefits their work with fertilizers had brought to UK farmers. Rothamsted Manor, in the center of the fields, has now become a boardinghouse for visiting scientists from around the world. This research station still studies various fertilizers but also looks at refining crops for energy production along with the long-term effects of pesticides, herbicides, and genetic modification.

The agricultural history of the last 160 years is written in the samples of soils, crops, fertilizers, and manures that the research station keeps in its "sample archive," where we see indications of increased production after the green revolution as well as evidence of the rise of pollution and fallout from Chernobyl in that same period. This portends poorly for man, because agricultural scientists hope soil will play a big part in doubling food production over the next several decades. We would need to double the amount of crops if we are to have enough grain on the table, feed in the barn, and biofuel in the tank in the future for man to keep going. A polluted and depleted soil report is not a healthy place from which to launch this increase in production, particularly as this may be only the first in a line of requests for more grain. The UN reports that we'll probably be pushing the limits of agricultural production into the next century.

To discuss the future of man and agriculture, we need to go back and take a closer look at the history of this relationship. At the end of the last ice age, about 12,000 years ago, as we entered the present interglacial period, the planet got warmer, rains fell more frequently, and plants grew bigger and faster than they had in over 100,000 years. Along the way, man realized that tending plants was easier than hunting game. The push in this direction may have come as man encountered lower populations of game or even extinctions of some key animals brought on by the development of human hunting skills.

The first evidence of domesticated wheat and barley appeared around 9500 BC, and shortly thereafter legumes such as lentils and peas. Farms were present first in the Fertile Crescent of western Asia. The idea quickly caught on, spreading to Egypt and India by 7000 BC, and gradually moved into Europe. Rice and millet started popping up in China around this time.

Man began to domesticate animals about the same time he domesticated plants. Goats were tamed around 10,000 BC in Iran and sheep around 9000 BC in Iraq. Cattle appeared around 6000 BC in India and in the Middle East. Agriculture spread more slowly over the northern and southern climates. It arrived even later among New World natives. Yet American Indians discovered maize and potatoes, some of the most important domestic plants in the world today.

Farming produced up to a hundred times more calories per acre than foraging, but it came at a cost to the health of new farmers. Hunter-gatherers rarely suffered from vitamin deficiencies, but farmers got scurvy, rickets, and beriberi because their diets were so base and unvaried. Infant mortality rose, also likely from poor diet. It seems that less protein, fewer vitamins, higher carbohydrates, and less movement were not what the doctor ordered. Humans who began to rely on agriculture shrank in height by almost five inches. Polynesians, American Indians, and Australian aborigines developed type 2 diabetes from their new high-carbohydrate diet and suffered a higher incidence of alcoholism. Alcohol consumption followed the growth of agriculture. There is some thought that barley was first domesticated for brewing beer rather than making bread. Tending crops, it appears, aroused a farmer's thirst.

Eventually, agriculture did result in larger populations, which led to the establishment of governments to protect and distribute grain, resulting in less fighting and longer lives. About nine thousand years ago the Sumerians invented counting tokens inscribed with pictures that could be impressed in clay to document land, grain, or cattle ownership. Scribes began drawing them with styli made of reeds. The result was known as cuneiform, perhaps our first written language.

Around one hundred thousand years ago, the world had approximately half a million people—counting *Homo sapiens*, Neanderthals, and other hominids. There were about six million *Homo sapiens* twelve thousand years ago at the end of the Ice Age. But then along came agriculture, and from 10,000 BC to AD 1, populations exploded about a hundredfold.

Agriculture improved life because it decreased competition for hunter-gatherers for a while, but then population growth caught up with the increased food supply. Greater populations, confined living, and proximity to domestic animals increased human contact with disease. Our impact on the environment grew. Wild animal diversity shrunk. While we were nomadic, our effect on the land wasn't too drastic, but once we settled down all hell broke loose.

THE POLLUTION PERIOD

Back at the Rothamsted Institute, I followed Kevin Coleman, a research scientist, into the institute's sample archive, a focal point for visiting scientists. Housed in a warehouse on the Rothamsted grounds are rows upon rows of five-liter bottles, all dated and stacked on shelves sixteen feet high that hold harvest grain, stalks, seed, and soil from test plots going back 160 years. On one high shelf is a sample of Rothamsted's first wheat field, dated 1843. To avoid mold, the bottles are all sealed with corks, paraffin, and lead. During World War II, samples were kept in discarded tins that once held powdered milk, coffee, syrup, and other wartime essentials.

The Rothamsted sample archive is a unique collection, since it comprises some three hundred thousand samples of crops and soils taken from agricultural field experiments for which the history is fully documented. "The samples are used by scientists worldwide to understand how changes in agricultural practices affect crop production, soil fertility, and biodiversity," says Coleman.

But what these vessels also contain is something researchers are not so proud of: a chronicle of human pollution. Over two centuries of industrial growth, soils have recorded what we've put into the atmosphere as well as what we've poured onto the ground. The Rothamsted sample archive holds evidence of nuclear atmospheric testing in Nevada and on the Bikini Atoll during the 1950s and 1960s. It also has a record of polychlorinated biphenyls (PCBs) from the manufacture of plastics and polycyclic aromatic hydrocarbons (PAHs) from power plants, fresh asphalt, and the fumes of automobiles. There are dioxins, the primary ingredient in Agent Orange, used to defoliate Vietnam. And plenty of heavy metals like zinc and copper from animal feed, cadmium from artificial fertilizers, chromium from tanning, and lead from pipes, vehicle fuel, industrial exhaust, and coal-fired power plants.

Many of these pollutants are deathly persistent. PCBs, the fluids that keep on lubricating and causing cancers, as well as DDT, the pesticide that keeps on killing, continue to appear in nature. Though amounts have gone down significantly since the 1970s, when both these chemicals were banned from most nations, PCB residuals continue to show up in the breast milk of Inuit mothers, and DDT continues to appear in freshwater fish and the raptors that eat them. DDT is still used in India to control malaria.

But the toxic residuals in our soils are something we may just have to live with. Right now, we have to get planting or starve.

THE NEXT GREEN REVOLUTION

That afternoon, out in the fields, Paul Poulton, a Rothamsted scientist, led me down rows of wheat stalks that displayed the results of a momentous moment in the evolution of agricultural products, the "green revolution." The seed heads were so thick that the plants appeared to be mostly seed, with a short, thick stock and little else. A light wind rippled through the rows in front of us, looking like ocean waves of wheat grain. The use of these new grains started shortly after World War II and spread like wildfire over much of the planet. Says Poulton, "Rothamsted switched over to these shorter, thick wheat plants about the same time the rest of the world did."

Norman Borlaug, an American agronomist, won the Nobel Peace Prize in 1970 for creating the first green revolution. A forester and plant pathologist, he walked away from a job at DuPont, a chemical company, in 1944 to join the Rockefeller Foundation's Mexican hunger project. His first post was as a genetics expert, but by the time he received his Nobel Prize in 1970, he was the director of the Wheat Improvement Program in Mexico.

Wheat was in poor shape in that country, the victim of a plague of maladies, including rust. Borlaug crossed Mexican wheat with rust-resistant varieties from elsewhere and obtained rust-resistance in wheat that grew well in the Mexican environment. Then he bred this wheat in the Sonoran Desert in winter and the Mexican central highlands in summer and developed breeds capable of growing in different climates.

Farmers in that country adopted the new varieties and wheat output began to climb. By the late 1940s, researchers knew they could induce higher grain yields with extra nitrogen, but the seed heads containing the wheat grains grew so heavy that the plants would topple, ruining the crop. So Borlaug worked at crossing wheat with strains that had shorter, thicker, more compact stocks. These plants could produce enormous heads of grain, yet their stiff, short bodies could support the weight without toppling. This transformation tripled and quadrupled production.

When researchers from India applied this idea to rice, the staple crop for nearly half the world, yields jumped several-fold compared with traditional varieties. Chinese agriculturalists started using semi-dwarf varieties to feed their people, a decision that aided China's rise to industrial power.

Now scientists tell us we need another green revolution if we are to meet the food demands of the next several decades. Our friends at Rothamsted are trying to participate in this, but it's not easy. Their current professed goal is to get twenty metric tons of wheat per hectare in twenty years, the so-called 20:20 Wheat. But Poulton says, "The average wheat grain yield for the UK is currently about 8.0 metric tons per hectare, but on the best soils with good management and favorable weather a farmer could hope to get 12 metric tons per hectare."

It seems the next jumps in crop production will come not from big discoveries like compact wheat but from a series of smaller changes that agronomists hope will add up to larger production. Rothamsted is currently looking at genetic improvements to increase the amount of grain; advanced pest and disease controls to protect plant yields; improved understanding of soil and root interactions to improve water and nutrient uptake; and a number of plant and environmental interactions to mitigate climate change.

Agricultural scientists at the institute are keeping an eye on what others across the Atlantic are doing as well. Jonathan Lynch, professor of plant nutrition at Penn State University, thinks that developing more aggressive root systems might be the answer to increased fertilizer efficiency and water usage. Crossing US beans with several varieties of ancestral stocks found in the high Andes Mountains, he's working to obtain belowground plant systems with lateral root reach sufficient to search for phosphorus in the topsoil and deeper taproots to go after receding groundwater and rapidly draining nitrogen.

Susan McCouch, professor of plant breeding and genetics at Cornell University, focuses on acid soils, a problem on 30 percent of the earth's surface. Acid releases aluminum into the ground, which inhibits root growth in plants, so the plants stop taking up water and nutrients, and they die. But McCouch is creating hybrid species of grains from ancestral lines, some in the wild, to achieve aluminum tolerance.

Researchers at Rothamsted are also working with the problem of acid soils by the use of biochar, what Brazilians refer to as *terra preta*, or black earth. Ancient Indian societies along the Amazon thrived using *terra preta*—charcoal from slow, smoldering fires—to enrich the relatively sterile tropical rain forest soils. Researchers are hoping that modern-day societies can do the same.

BLACK EARTH IN THE AMAZON

To get a picture of the potential of *terra preta*, one must visit the central Amazon near Manaus. I flew into Venezuela in early August and took a two-day ride aboard a bus, which climbed up and over the Sierra de Pacaraima down into the Amazon Basin. The road wound through the forested mountains in the dark night, and the way looked clear until around the bend ahead came another bus. At the last second, both buses veered toward the outer shoulders, and as we whisked by each other dangerously, a hanging limb from the jungle struck our right front window and turned it into a giant spiderweb of glass, which the driver chose to ignore. His side of the double window was still clear.

We traveled all day, first through savanna, then dense tropical forest, and arrived by evening in the city of Manaus perched at the junction of the Rio Negro and the Amazon River. The city was alive with vendors, farmers, and tourists in the afternoon sun. Manaus is the largest city in the central Amazon. A group of archaeologists greeted me at the station, and soon I was headed by ferry across the Rio Negro to their field site on the Amazon.

By morning we rolled out of hammocks, ate a hearty breakfast of eggs, fruit, bread, and coffee, then headed out to the field. University of São Paulo archaeologist Eduardo Góes Neves and some fifty volunteer archaeologists from Latin America, the US, and the UK were excavating an archeological site on a papaya farm that overlooked the Amazon River. This location harbored community graves and other ancient relics going back more than two thousand years. The lush orange color of the fruit and the robust green leaves on the trees were due to the soils left by ancient Indians who once occupied these lands.

The banks of the river were plentiful with *terra preta*, a gift of civilizations past. While most Amazonian soils were notoriously nutrient-poor, yellowish, and sterile, *terra preta* was dark, fragrant, and rich—a farmer's delight. Neves and others believe that by devising a way to enrich the soil, early inhabitants created a foundation for agriculture-based communities that harbored far greater populations than was previously imagined.

Amazonian soils have very little rock in them, which means that early civilizations made their homes and worship sites from wood. These structures, no matter how elaborate, degraded over time, leaving little evidence of past human glory. The principle evidence of ancient civilizations was in the ceramics they formed and fired, pieces of which have survived in the soil.

Early Amazonian life wasn't easy. Indians used stone axes to fell trees along the banks of the rivers. The task was long and tedious, taking days to weeks to cut down large trees. The process created small openings in the forests, letting in some light, but not enough to thoroughly dry out the vegetation. Farmers started fires to clear the forest for crops, and the fires would smolder for days, creating charcoal that was the basis of *terra preta*.

Most Amazonians today use "slash-and-burn" methods to create space for their crops. Natives use chain saws to clear much larger spaces than the ancient Indians did. This creates large spaces with lots of light, plentiful kindling, and huge, hot fires that produce quantities of ash but little charcoal. Ash has sufficient nutrients to last a few seasons, after which the land goes fallow, whereas *terra preta*, or biochar, can last far longer. One farmer near Neves's study site cultivated crops on *terra preta* soil for forty years without ever adding fertilizer. William Woods, a soil scientist and professor of geography at the University of Kansas, claimed this was amazing and told me: "We don't even get that in Kansas," a US state famous for its soil.

Rothamsted Research has tested soils at several sites in the Amazon and found that *terra preta* took up, integrated, and retained carbon from organic matter much more freely than typical native soils, and this was one of the reasons for increased yields.

Perhaps this would stave off starvation.

THE KING COTTON FIASCO

On another trip much farther north, but still looking at dirt, I traveled with Dan Richter, professor of soils and forest ecology at Duke University, and several of his faithful graduate students in a caravan of cars heading south from Durham, North Carolina, and the Duke campus to the Calhoun Experimental Forest, one of Richter's favorite field sites. Located in Sumter National Forest near Union, South Carolina, the Calhoun was established in the 1940s to study the serious problems the region had with its soils.

Richter worked in the Piedmont area of the southeastern United States. Here, years of Southern cotton production on farms and plantations in the 1800s had eroded earth, extracted vital nutrients, and greatly diminished the productivity of the region's soils and ecosystems.

The Calhoun's initial location was picked to represent the "poorest Piedmont conditions" of agricultural soil erosion and cropland abandonment. Early studies on the Calhoun were aimed at soil improvement and watershed restoration in order to find the cheapest, quickest, most effective ways to improve tree growth and soil structure and to increase soil fertility for plants. Duke University's long-running collaborative study with the US Forest Service has aimed at monitoring, sampling, and archiving information.

The Piedmont is a plateau region of the Eastern United States between the Blue Ridge Mountains, the eastern range of the Appalachian Mountains, and the Atlantic Coastal Plain. It stretches from the state of New Jersey in the north to Alabama in the south. It is an area of about 80,000 square miles (210,000 square kilometers), and the soils in this region are predominantly clay and moderately fertile. In the central area of North Carolina and Virginia, tobacco is the main crop, while in the north, the focus is on orchards and dairy farming. In the south, where cotton was the chief crop in the 1800s and early 1900s, all that is left, said Richter, "is one of the nation's most degraded landscapes."

He took us to a place in the forest where the soil had been excavated, exposing its profile. In the caravan there were a number of Asian students, including several from China who were eager to absorb the lessons and apply them to similar problems in their own country. China has long-term soil plots that are twenty and thirty years old "and they test soil and crop changes in a wide variety of soils to the major agricultural inputs, organic and inorganic," said Richter.

He noted that only 150 years of Southern cotton production had caused the erosion of as many as eight inches of topsoil across the southern Piedmont. Native forest was coming back here as it has in the northeastern part of the US, and this was currently restoring some of the organic matter, but the region was still deficient in many vital nutrients, including nitrogen and phosphorus.

These are still agriculture's two most important fertilizers. Though most crops need a lot less phosphorus than they do nitrogen, the smaller amount of phosphorus is critical to the mixture if you want to get robust growth. Nitrogen can be produced artificially, but phosphorus has to be mined, and the process is ugly.

West-central Florida produces much of the US phosphorus and most of that is taken along the Bone Valley Member of the Peace River Formation, one of the richest sources of the mineral in the world. Miners create what looks like iridescent volcanoes from above as they pull phosphorus ore out of the ground, crush it, and process it into acid lakes. There, the ore is separated into gypsum, which forms the volcano-like cone and the fluorescent green acidic liquid. Farmers use that liquid in a dry mineral form to feed their crops.

The trouble with phosphorus mining is the occasional breaking of these cones, which can spill into the Peace and Alafia Rivers, delivering the spillage to Charlotte Harbor or Tampa Bay. Here, phosphorus contributes to algae blooms, which depletes the oxygen in the water and asphyxiates the fish.

"If you inventoried the world's reserves of fertilizers, you'd find that phosphorus is by far the most limited in supply," said Richter. "By some estimates of known deposits we have maybe fifty to one hundred years of it left. Lots of ecosystems are limited by phosphorus, yet lots of our surface waters are currently polluted by it. It is second only to nitrogen in the amount applied to agricultural lands, and it is a critical resource for corn, cotton, rice, wheat, and other grains. Without it, these crops won't grow to their potential."

Still, nitrogen is the biggest problem right now, not the shortage of it, but its overwhelming presence in all of our soils. William Schlesinger at the Cary Institute of Ecosystem Studies told me, "Since World War II, synthetic production has doubled the amount of nitrogen on the planet, and that has seriously changed the chemical environment that a lot of species have evolved in."

Nitrogen is created by the Haber-Bosch process, developed by German scientists before World War I, which utilizes high temperatures and very high pressures along with various catalysts to produce ammonia (a common inorganic fertilizer) from nitrogen and hydrogen gas. The process uses heavy amounts of electricity and is potentially poisonous to the environment.

Gaseous emissions of nitrogen can drift with the wind, landing on fields and promoting some species of plants while eliminating others, again reducing the number of species. This happens in farms in the Mississippi River Valley, from which nitrogen emissions drift northeast. According to Schlesinger, this produces a type of acid rain similar to the acid rains of the 1980s and '90s that resulted primarily from sulfur emissions from power plants mixed with rain to create sulfuric acid. The current problem stems from nitrogen emissions mixed with rain that create nitric acid.

Duke biologist Richter currently leads a project with Rothamsted to compare soils augmented by inorganic nitrogen to soils augmented with organic manure. High nitrogen from inorganic fertilizers can acidify soils far more than acid rain, and the cumulative effects can be disastrous.

Back on the campus of Duke University, Richter slid open one of a tall stack of drawers in a chest just outside his office, a proud smile on his face. The drawer was packed with quart-size jars, each filled with soil from the Calhoun Experimental Forest in South Carolina. It was not nearly as large as the Rothamsted sample archive, but it nevertheless represented the careful work of students and researchers from his department and the US Forest Service who are looking for changes in the soil over sixty years, the length of Duke's study of Calhoun soil. This study is part of a database of long-term soil studies that Richter is trying to establish across the globe.

Richter showed me a sample from the year 1963, which he held up to the light, brandishing its chestnut brown color. "Nineteen sixty-two was the year Khrushchev and Kennedy were negotiating the atomic test ban treaty to halt atmospheric testing of nuclear weapons, and in those years you see the highest concentration of carbon 14, a signature of radioactivity in the earth's soil," said Richter.

"Bomb carbon," as he called it, had originally been sited in a series of wine bottles that were hermetically sealed. The grapes used to make the wine had tasted the radioactive air during photosynthesis and held on to it during the winemaking process. According to Richter, it was not directly harmful, but it did give scientists a small "slug of radioactivity" which they could use to determine how soils built and changed organic matter over time.

NEW SPACES

A week later, Angelica Pasqualini, a stylish Italian research assistant at Columbia University, took me, her computer, plus two extra sets of shoes in her fashionable bag to the rooftop of Regis High School, an all-boys Catholic school on East Eighty-fourth Street in Manhattan, to show how the school was promoting the use of rooftops for agricultural space. Trays of drought-tolerant plants on top of the building provided natural insulation that reduced heating and air-conditioning costs for the school. If the city's one billion square feet of roof space were transformed into green roofs, it would be possible to keep more than ten billion gallons of water out of the city sewer system. "Green roofs help retain storm water runoff," said Pasqualini. They also provide cheap insulation from the weather and a place for the bees.

In a corner of the same roof a local beekeeper, Joanne Thomas, checked her beehives to see how her flower breeders were doing. It seems that Manhattan roofs are going green. Her bees stayed healthy on pollen from uptown rooftop blossoms. They also pollinated local plants up and down the street as well as in nearby Central Park.

On the roof of a gritty industrial building in Brooklyn, a small urban farm called Gotham Greens provides a leafy oasis of green leaf, red leaf, and baby butterhead lettuces, plus Swiss chard, Chinese cabbage, arugula, and basil. Turning the rooftops of New York into mini farms could provide more open earthen space, a commodity that is rapidly diminishing.

Dickson Despommier, a Columbia University professor and author of *The Vertical Farm: Feeding the World in the 21st Century*, told me later in his office that abandoned skyscrapers were a viable alternative to open space in New York and other cities throughout the world. This was particularly true in cities in the American Midwest, where urban flight and a struggling economy have left many tall buildings vacant. According to Despommier, these structures could work as hothouses—protecting crops from weather, providing ample window light, and even allowing access to elevators to haul crops up and down for planting and harvesting.

Grains, fruits, and vegetables are an important part of the future agricultural picture, but what about meat? It takes five times the amount of grain consumed by the entire American population to feed the seven billion domestic livestock animals in the United States. In other words there is a cow, pig, lamb, or goat out there for every

American. Only, they eat five times as much as we do. If we ate a little more grain and a little less meat, it would go a long way toward solving our looming food shortage. But grain shortages are not the only problems livestock create.

In 2006, the United Nations Food and Agricultural Organization report *Livestock's Long Shadow: Environmental Issues and Options* highlighted that cattle-rearing generates more greenhouse gases than driving cars. It is also a major source of land and water degradation. If emissions from land use are included, then livestock accounts for 9 percent of CO_2 derived from human activities. It also generates 65 percent of human-related nitrous oxide and 37 percent of methane, both of which are much stronger greenhouse gases than CO_2. And it accounts for 64 percent of ammonia, which is a heavy contributor to acid rain. Livestock now uses 30 percent of the earth's entire land surface and is a major cause of deforestation in South America.

But as countries develop and standards rise, people want to try meat, a symbolic reward for joining the middle class. Yet if, in the process of gaining knowledge and riches, their appetites turn to beef, it could negate the advantage of putting off childbirth in developing countries.

ENDANGERED SOILS

Man's greatest challenge may be finding enough land: we are running out of it. We won't have enough to grow grain or raise enough cattle if populations don't stabilize. Ronald Amundson, professor of soil science at the University of California, Berkeley, in agreement with Duke's Richter, thinks that soils are critical components of the earth's biosphere, but they are being rapidly transformed by agriculture and urbanization. It's not a small matter. "The combined human impact on land surfaces during the past few hundred years is as large as that which occurred during the last ice age," says Amundson. Soil types depend on a combination of climate, geology, and topography to help form them. There are twenty thousand soil types in the US and their natural acreage is now severely depleted. Amundson uses a value of 50 percent reduction to indicate endangered and 90 percent reduction to indicate extinct. "Many types of soil in the United States are currently endangered and a handful extinct," he wrote to me.

Some scientists speak of an end to population growth as developing nations send their citizens to school. But many think that our numbers will continue to mushroom longer.

According to UC Berkeley's Anthony Barnosky, "With seven billion people alive today, we are devoting 43 percent of earth's land to agricultural production in one way or another. By the time we hit eight billion we are going to be up to 50 percent of earth's lands devoted to agriculture. By the time we hit nine billion it will be more like 60 percent of available land. And remember, not all land is created equal. We are already using the best 43 percent. There would be some huge problems."

A great number of wars have been fought over food and the space to grow it. The Revolutionary War in the US started with a dispute with the British over tea. Around this time, the British fought with the French over control of sugar-rich Jamaica. The Guerra del Sale (the War of Salt) in the late fifteenth century was fought between Venice and the northern Italian city of Ferrara over the right to tax salt.

In Jared Diamond's book *Collapse: How Societies Choose to Fail or Succeed*, he argues that the genocide and murder of 800,000 Rwandans in 1994 was fueled not solely by ethnic hatred but also by a population too big to feed itself. Their lands had been divided and redivided so many times and grown so small that the remaining plots could no longer feed their owners. We could face similar situations in the future.

Starvation weakens the body and the spirit. The landscapes we are changing and the species we are killing off will produce ruinous effects in our fight against disease, the next of our growing challenges.

6

WARNING SIGN II: OUR BODIES

As our agricultural practices decrease the biodiversity of our plants and animals, one of the unintended consequences is the increased presence of disease. Over the last half century a number of new diseases have developed in our world and we are only beginning to understand our role in their development. The loss of native species decreases the dilution effect that results from having a variety of carriers of disease, some more efficient than others, the less efficient ones decreasing the threat from new diseases. At the other end of the equation, we find that our animal husbandry practices are decreasing our ability to treat disease by creating antibiotic resistant germs.

A prime example of this threat can be found in the story of Mr. Yu. G. (health authorities used initials to indicate him), a storekeeper who lived in a cotton factory in the town of Nzara in southern Sudan in the late 1970s. He was a quiet man, a recluse. He worked at a desk in the back of a cotton factory with cloth piled all around him. Bats roosted in the ceiling near his desk, which many suspected of making him sick, but no one was able to prove a connection.

Still, Mr. Yu. G. became internationally famous on July 6, 1976, when he went into shock, blood pouring from all his orifices, and died. He never made it to the hospital. He was the first case ever, the index case, of Ebola Sudan. Mr. Yu. G. shared his office with two other workers who got the disease a few days after his death. They, too, went into shock, began to hemorrhage, and died.

One of the dead coworkers, P.G., was unfortunately more social than Mr. Yu. G. He had friends and even a few mistresses. The disease spread from him more rapidly. It went through sixteen generations of infection, killing many of the hosts, raging through the town of Nzara and eastward to the town of Maridi, where there was a hospital. It then went through the hospital, jumping from bed to bed, finally going after the medical staff; some got sick, and others grew frightened and fled the hospital. The World Health Organization (WHO) sent a team of investigators, which found that the medical staff's departure actually turned out to be a blessing, since they had been using dirty needles to inject their patients, unknowingly spreading the disease. Once the practice had stopped, the spread of the disease subsided.

The death rate for Ebola has been high. Of the 284 people who got Ebola hemorrhagic fever in Sudan in 1976, 151 (53 percent) died. Of the 318 people who got Ebola in that same year in Zaire, 280 (88 percent) died. Of the 264 people infected in a more recent outbreak in Zaire in 2007, 187 (71 percent) died. The disease was detected in Uganda and the Democratic Republic of the Congo in 2012. Since the disease travels by blood, it is not as contagious as a common cold, which is spread through the air. It kills so quickly that victims often don't have a chance to spread it. They're simply not around long enough to come into bodily contact with many others.

Recently Ebola has again reared its ugly head and is raging through West Africa. In Kenema, Sierra Leone, the government is trying to quarantine victims in the local hospital. But so many patients and health workers have died there that some victims simply get up and go home, deserting the place because they feel it's a death trap. In doing so, however, they're only promoting the spread of the disease and undermining international efforts to contain it.

Some infected health workers and missionaries have been flown back to their home countries, but are kept in isolation. As we were going to press, the toll was rising into the thousands, the most serious outbreak yet. It's a small number compared with deaths in Africa from malaria and HIV/AIDS, but it is another reminder of how serious disease can suddenly reappear.

Some scientists believe that the disease may have been brought to humans from monkey meat. The virus first erupted in a population of research monkeys sent to the city of Marburg, Germany, in 1967, at a place called the Behring Works, which used African green monkeys to produce vaccines. Some of those monkeys turned out to have come from a group of islands on Lake Victoria. Epidemiologists believe that AIDS may have also come from monkeys on these same shores. According to Richard S. Ostfeld, a disease ecologist at the Cary Institute of Ecosystem Studies, about 60 percent of all infectious diseases that affect humans are "zoonotic," meaning they reside in animals that act as reservoirs for the disease. But this animal association increases to 75 percent for new and emerging human diseases.

The risk of disease outbreak grows when we disturb natural habitats and reduce the biodiversity of the land. "Bushmeat hunting is clearly responsible for the initial outbreaks of HIV/AIDS, Ebola, and many other viruses," Ostfeld told me. As animal numbers go down, the animals that are left harbor the most disease. "The principles seem the same: the best wildlife reservoirs for the pathogen are also the species that thrive when biodiversity is lost."

The risk of a number of these horrible killer diseases, including Ebola, severe acute respiratory syndrome (SARS), Middle East respiratory syndrome (MERS), and others, are linked to environmental destruction. Ostfeld studies Lyme disease, for which New England rodent populations are a critical part of the disease cycle, just as they are in monkey pox, hantavirus, and tick-borne encephalitis.

"Thirty years ago, these diseases were absent from our landscape," said Ostfeld. "Now they are established and spreading. When humans fragment the habitat and reduce species diversity, the probability of catching these diseases increases." He believes that studying the ecosystem of these diseases can give us better insights to the ecosystems of others.

A broader range of different animal species means that the effects of the disease are spread out and diluted. With more animal species there are more hosts for the disease, and some of the hosts are going to be less effective at passing the disease on, thus diluting its total effect. Biodiversity also allows for more predators, and that can reduce disease host populations.

With respect to controlling critical outbreaks, Ostfeld thinks that scientists sometimes rush to judgment and make inappropriate decisions. He cites the response to SARS, a serious form of pneumonia, as a typical misguided counterattack against an infectious disease. In 2003, WHO physician Carlo Urbani first identified SARS in a forty-eight-year-old businessman who had traveled from China to Vietnam by way of Hong Kong. The disease had started in China's Guangdong Province and spread from there. The businessman was admitted to the French hospital in Hanoi, worsened, and died. Dr. Urbani died from the disease just weeks after he helped to identify it and warned the world of its dangers. SARS infected more than eight thousand people and killed 774 around the world before it was brought under control.

When SARS first broke out, scientists quickly identified it as a virus probably transmitted to humans through an animal. Researchers from the University of Hong Kong examined twenty-five animals from eight species in a live animal market in southern China, and found a SARS-like virus in all six civets they sampled, as well as in a badger and a raccoon. A civet is a small, catlike animal native to the tropical forests of Asia and Africa. It has a pointed snout like an otter. The term "civet" applies to over a dozen different mammal species.

Authorities quickly rounded up and destroyed all the civets sold in these markets. Ostfeld told me, "Civets weren't the real culprit; it was fruit bats. The bats may have contaminated an area used by civets with bat urine and feces, much like a dog would, but it is highly unlikely the civets were the source of transmission to humans." Two different studies of SARS confirm that bats were the real reservoir of the disease.

Another example of misidentification occurred when health authorities went after bovine tuberculosis, an illness that affects cattle and is a tremendous financial risk to meat and milk producers. Health authorities found that one of the ways for cattle to get exposed is through contact with badgers. In Europe and the UK, studies show that badgers are the natural reservoir. So officials started killing badgers. "Only they found that badgers are an intensely social animal that stay together," says Ostfeld. "But if you disrupt their environment, they will disperse, start to run around more, and in the long run increase exposure to cows."

According to Ostfeld, when a disease outbreak hits, first responders at government agencies are good at mobilizing quickly and figuring out what the pathogen is. But when they try to figure out where the disease comes from, they aren't as capable. He thinks that it's not enough to identify the most obvious players—the pathogen and a host or two. These may be only parts of a greater cast of characters. Without an understanding of the complete ecology of the disease, some health responses may enable a wider spread of the disease. Fruit bats are reservoirs of Hendra virus, an acute respiratory and neurologic disease found in horses and humans in Hendra, a suburb of Brisbane, Australia. And the response has been to go in and cut down the trees to drive out or kill the bats. But when bats don't have enough to eat or are disturbed, their immune response lowers and they tend to shed a lot more virus. This is an example of the consequences we face when we change the natural landscape to fulfill a purpose different from the one for which it originally evolved.

CRITICAL MASS AND CROWD DISEASES

The advent of agriculture has disrupted the natural processes of evolution. As we discussed before, it changed life for man by increasing food production, but it also increased the presence of infectious disease. As man grew more food, he was able to have larger families, and with more people came more garbage and sewage. The increase in farm animals drew more rats and mice, creatures that brought a number of serious diseases, including typhus and the bubonic plague.

It is widely understood that you need a certain number of people in close proximity for disease to spread. The critical mass—the point at which the disease achieves optimal virulence and transmissibility—for measles is a half million. Measles could not have raged in the days before agriculture, when man lived in small groups and hunted for a living. Chicken pox had an easier time making the transition from hunter to farmer, since its critical mass is only about a hundred.

At first, infectious diseases were a much bigger problem for farmers than for hunter-gatherers. But slowly evolution selected for the farmers who had better immune responses to these outbreaks. These Old World immune responses were also passed on to city dwellers when the farmers went to more densely populated areas to sell their goods. All the while hunter-gatherers in the Americas and Africa developed fewer immune responses to what were basically crowd diseases; they didn't have as many crowds.

Some people in Africa even managed to develop immunities that protected them from malaria, which is not a crowd disease. It is transmitted among humans by female mosquitoes of the genus *Anopheles* during the blood meal they must take to produce eggs. While this may seem a blessing for people in Africa, immunity to malaria can come at great cost. One of the best known is sickle-cell anemia, which can itself be a very serious disease. About 250,000 children are born each year with this disease. Sickle-cell anemia is an inherited disorder in which red blood cells are abnormally shaped and may get stuck in blood vessels, making the delivery of oxygen throughout the body difficult. It is a chronic disease, though it can be treated. It strikes Africans more than others, though the trade-off is their immunity to falciparum malaria, a response that scientists have been trying to accomplish for hundreds of years. According to the World Health Organization, malaria in 2012 caused an estimated 627,000 deaths, mostly among African children.

There were no vaccines available when Columbus and other explorers brought the Old and New Worlds together, and it was the New World that was the least prepared. American Indians as well as Australian aborigines, Polynesians, and many island populations had never encountered the strange crowd diseases that arrived with the Old World invaders, and against which the natives had few natural defenses.

American Indians had migrated from Northeast Asia to the Americas about fifteen thousand years ago, before the advent of agriculture and crowd diseases. Thus they had no resistance, and the place they went to, the Americas, was so sparsely populated that they couldn't grow immune to crowd diseases. These first American travelers had to pass through Siberia and Alaska to get to their destination, and they left behind tropical insect-borne diseases such as malaria.

Michael Greger, MD, author of *Bird Flu: A Virus of Our Own Hatching*, believes that the real reason Old World diseases like smallpox never developed in the Americas prior to the conquistadors was that there were far fewer domesticated animals in the New World. The last ice age and its hunters had knocked off most of the easily domesticated animals like American camels and horses, leaving the indigenous population to raise for food species such as llamas and guinea pigs, none of which were good carriers of lethal human diseases.

The dramatic differences that these selection pressures brought to the table was evident when Old World explorers started coming into regular contact with New World natives. European diseases such as smallpox, whooping cough, measles, diphtheria, leprosy, and bubonic plague attacked the unprepared immune systems of American Indians, with devastating results.

In the tropics, malaria and yellow fever joined the list of infectious agents, and native populations dropped by 90 percent or more, according to some estimates, in just a few centuries. Hernán Cortés conquered the Aztecs, and Francisco Pizarro conquered the Incas, aided by the introduction of Old World diseases.

FOLLOWING THE EUROPEANS

Disease followed Western Europeans into the Amazon as well. In 1542, Spanish explorer Francisco de Orellana and his men headed down the east face of the Andes and on to the Amazon River looking for El Dorado, the mythic "City of Gold." His expedition found villages, towns, and well-developed societies with agriculture, ceremonies, and elaborate wooden structures. They reported passing twenty villages in one day, and one settlement that "stretched for five leagues"—a league being the distance a person or a horse could walk in one hour.

Still, the lack of gold, the hostility of the tribes (who often greeted the Spanish boats with fusillades of poison arrows), and the treachery of the Amazon River itself left Orellana's expedition in rags. By the time further explorations were mounted, the dense populations Orellana had witnessed were gone. Some five million people may have lived in the Amazon region in 1500, but by 1900 the population had fallen to one million and by the early 1980s it was less than two hundred thousand. Recent archaeological evidence supports Orellana's accounts of dense populations. Scientists believe that disease, perhaps even arriving with Orellana's expedition, spread throughout the Amazon, devastating ancient cultures as it had elsewhere.

If American Indians had not succumbed to Old World epidemics, they would have been better able to adapt to European military strategies, and the going would have been far more difficult for the conquerors.

This tendency to be felled by Old World diseases upon initial introduction continued in South America even into modern times. In 1967, a missionary's two-year-old daughter came down with measles in a village of predominantly Yanomami Indians in Brazil near the northern border with Venezuela. Almost all of the 150 Indians, young and old, caught the disease and one in ten died, despite the desperate efforts of the missionaries.

First contacts were often the most deadly, killing one-third to one-half of the native New World population in the first five years. In Brazil, of eight hundred Suruí Indians who were contacted in 1980, six hundred died by 1986, mostly from tuberculosis. As Charles Darwin said, "Wherever the European has trod, death seems to pursue the aborigine."

ON TO AFRICA

However, when Western Europeans tried to conquer sub-Saharan Africa, disease went after the explorers, not the aborigines. Europe didn't make serious attempts to explore Africa until the fifteenth century, about the same time they discovered America. The king of Portugal sent eight men, one of the first expeditions, up the Gambia River around 1500 but only one came back alive.

Europeans bought slaves at coastal outposts or on offshore islands. Going deeper into the jungle posed too many dangers from native ambushes, poisonous snakes, and disease. British soldiers stationed on the Gold Coast might lose half their men in as little as a year. Arab or part-African slave traders seemed to be less susceptible. Mungo Park, a Scottish explorer, made his second attempt to explore Africa in 1805 with a party of forty-five Europeans. Only eleven still stood when they reached the Niger. Dr. David Livingstone, the famous Scottish medical missionary, lasted for a while but eventually fell to malaria, as did his wife.

In the early twentieth century, quinine, a drug developed from the cinchona shrub of South America, became available as an antidote for malaria. At the same time efforts to control mosquitoes helped prevent the spread of this disease and yellow fever, as did efforts to control the tsetse fly, which caused sleeping sickness. With these potential killers held at bay, European countries ventured into Africa and quickly conquered almost the entire continent.

Africa did not become another America. Europeans did not displace

Africans. It seemed that in order for Europeans to take command, the locals had to die off, and Africans did not die off as American Indians did. Tropical diseases bested anything the Europeans brought with them. The Africans had some selective resistance to these diseases, whereas the Europeans had none.

Resistance is particularly unreliable with new diseases. Back at the Cary Institute, biologist Rick Ostfeld stood in a New England forest with a white-footed mouse in his hand. Both Ostfeld and I were dressed in white coveralls with latex gloves. A very-much-alive mouse had been collected from a trap the previous night. Ostfeld ran his index finger through the mouse's back fur, spotting several ticks attached to its skin. It was spring and the forest around us was thick with tall trees, bright green leaves, and lots of ticks. The biologist carefully pulled one of the ticks off the mouse and showed it to me. "If that tick bit you, you would have a 40 to 45 percent chance of catching Lyme disease," he said. I stepped back.

Ostfeld, a senior scientist at the institute—a calm, meticulous man whose broad shoulders attest to his weight lifting—has been studying Lyme disease here for twenty-four years. Dutchess County and four other mid–Hudson Valley counties have the nation's highest rates of Lyme disease.

Ostfeld and others at the Cary Institute are investigating the ecosystem surrounding Lyme disease, the West Nile virus, and similar diseases recently proliferating in the US that are transmitted by ticks and insects through their animal hosts. Recently they discovered that the black-legged ticks that spread Lyme disease could also infect people with Powassan virus encephalitis, which can cause central nervous system disruption, meningitis, and even death in 10 to 15 percent of reported cases. Adding to the problem is the fact that unlike Lyme disease and other ailments carried by black-legged ticks that take hours to transmit once a tick is attached to its victim, Powassan virus encephalitis and its variants can be transmitted in just fifteen minutes.

This leaves very little "grace period" for removing ticks, and underscores the importance of vigilance when in tick habitats.

Tick removal has become a critical activity of late for Northeastern outdoorsmen. Lyme disease was first reported in the United States in the town of Lyme, Connecticut, in 1975. "It reached a high of 30,841 reported cases in 2012," says Ostfeld. "But the CDC [Centers for Disease Control] has recently estimated that reported cases represent only 10 percent of the actual number of cases, so it is likely that Lyme disease exceeds 300,000 cases per year." Such statistics keep some hikers at home on the couch on weekends, a not-too-healthy alternative.

Most cases occur in the Northeast and upper Midwest, but there have been many reported along the Pacific Coast and elsewhere. If diagnosed in the early stages, Lyme disease can be cured with antibiotics. It may start out feeling just like the flu, which is sometimes ignored by patients, but the results can be severe. Without treatment, the CDC claims it can affect joints, the heart, and the nervous system—causing pain, paralyzed facial muscles, and nerve damage in the arms and legs.

In his laboratory at the Cary Institute, Ostfeld showed me a slide of the slender spiral-shaped bodies of *Borrelia burgdorferi*, the bacterium that causes Lyme disease and all its symptoms. Certain ticks carry the bacterium, though they aren't born with it. They acquire it when they bite an infected mouse or chipmunk. Man acquires the disease when bitten by an infected tick.

The black-legged tick (*Ixodes scapularis*) carries Lyme disease bacteria in these woods. It goes through three stages in its short two-year life span—as larvae, nymph, and then adult—with each period requiring at least one good blood meal before moving on to the next. It is during these blood meals that the ticks acquire the disease and pass it on.

Ostfeld came to the Cary Institute in 1990 with a background in the behavior and evolutionary ecology of small mammals like voles, which undergo periodic, dramatic population swings. The biologist studied the disease, black-legged ticks, the ticks' animal hosts, and the forest that surrounded them to see how all the players in this disease drama functioned together. Ostfeld and his colleagues soon realized that the booms and busts he witnessed in small mammals and in forests themselves could play an important role in the spread of infectious disease.

The cycle may begin with an abundant crop of oak acorns in any year. Because acorns are a highly nutritious and long-lasting food source, they create an explosion of white-footed mice and eastern chipmunks in the following year. These small mammals are the preferred hosts of black-legged ticks. Still, the ticks must go through several phases before they start transmitting the disease to man, meaning the risk of Lyme disease is highest two years after plentiful acorns. It's a complex system.

It is understandable that people in New England, the Middle Atlantic states, and the upper Midwest live in fear of contracting Lyme disease, but many use it as an excuse to stay out of the woods. Press reports of the disease have people believing that ticks are much more abundant than they once were, and that Lyme disease is spread by ticks carried by deer. This belief has resulted in calls to dramatically reduce deer populations in different areas of the Northeast. Black-legged ticks are sometimes called deer ticks, though Ostfeld claims that deer are not such important carriers as rodents are.

Ostfeld found that when deer are reduced by hunting or excluded by fencing, disease rates actually increase over the next few years. That's because deer are highly unlikely to transmit a spirochete infection to feeding ticks; they are good hosts for ticks but not for the disease. Small mammals are much better at handing off infections to ticks. Thus deer protect people from Lyme disease by being lousy hosts for Lyme-bearing ticks, "so taking away deer, at least initially, removes the protective role they play in reducing tick infections," said Ostfeld.

He told me that ecologists tend to be excluded from the pool of rapid emergency funding, and are often left out of the first-response teams when new diseases appear. Also, it seems that money is more available for what Ostfeld explained as "the disease of the month." Funding for SARS and West Nile peaked in the year or two after the worst disease outbreaks. "Ironically it is the study of those well-established diseases that give us a good grip on how disease systems work. We shouldn't abandon these intensive studies. They are the gold mine from which we get an understanding of basic disease processes," said Ostfeld.

In the course of Ostfeld's studies, he has learned a number of things about Lyme disease. He knows that taking a walk in a fragmented forest, one split up by roads and development, is more dangerous than taking a walk in extensive virgin forest. And he knows that the more opossums, squirrels, and foxes there are in a forest, the less chance there is of catching Lyme—and he suspects that the same is true for the presence of hawks, owls, and weasels. He is focusing now on determining the reasons for these observations, a major part of his studies.

As we've said, forest fragmentation enhances the spread of disease. Fragmentation occurs when large, continuous forests are divided into smaller pieces, either by roads, agriculture, urbanization, or other human development—shrinking the area available to animals and plants that rely on the habitat. Some critters, like predators and large-bodied animals, need large areas to maintain viable populations. Some are poor dispersers for whom the strip mall or suburban development is a severe barrier. The results are species losses. The species that are most resilient to fragmentation—mice, chipmunks, etc.—are often the only ones that remain. And these are the bad guys when it comes to disease transmission.

Ostfeld told me that if the tick we found on the mouse earlier that day had bit me, my chances of catching Lyme disease would have been at least 40 percent, but if we'd found the tick in a vacant woodlot in the nearby town of Poughkeepsie, my chances would be closer to 70 or 80 percent. A woodlot is an example of fragmentation, and the town of Poughkeepsie has lots of that.

To test Ostfeld's theory that fragmented forests increase disease, he and a number of biologists selected fourteen forest fragments that were similar in types of vegetation but were isolated from other suitable habitat for Lyme hosts. What they found was that the larger the forest patch, the smaller the proportion of black-legged ticks infected by the disease it contained.

In the Midwest, tracts of trees in the middle of corn and soybean act like islands in the middle of the ocean. Corn and soybeans play the part of the ocean, since they are sufficiently inhospitable to native animals and create a barrier to dispersal much like the ocean would to an island animal. Thus corn and soybean crops deter many larger forest animals, but small mammals like mice and chipmunks do just fine there. Wooded lots have fewer species altogether, but the animals they do have are the disease amplifiers. Ostfeld found that the larger the size of the wooded tract, the smaller the proportion of diseased black-legged ticks there. But in smaller lots the numbers of infected ticks are astronomical.

The idea that peaceful patches of forest amid cornfields could be harbingers of disease is ominous, but so is the fact that the antibiotics we've been counting on to treat those diseases may not be able to help us for much longer.

THE RISE OF SUPERBUGS

Antibiotic resistance is growing so fast that we may soon have nothing left to tackle the new diseases we are fostering. The failure of our medicines is due to farmers who use antibiotics in animal feed to fight off diseases promoted by overcrowded conditions in confined-animal feeding operations for pigs, chickens, and cattle. These practices are creating superbugs that are immune to the antibiotics they've already adapted to.

Resistance to antibiotics normally occurs if your doctor prescribes a dosage that is not sufficient to eliminate the disease, or if you don't take the full number of doses prescribed. In the process, the disease gets stronger and is less affected by the medication on subsequent usage. Bacteria that survive the first treatment multiply and are resistant to the next treatment.

But antibiotic resistance can also come from eating meat from animals once treated by antibiotics. Disease is a particular problem in confined feeding operations where animals are kept in close quarters and fattened up for market. Putting antibiotics into animal feed is meant to lessen the threat of disease and to promote animal growth, but some scientists are finding that part of our increased resistance to antibiotics comes from eating animal products tainted by antibiotics.

Feeding our cattle, chickens, and pigs low doses of antibiotics is a setup for our own resistance to the medicine. In reality, low-dose medication that is not monitored selects for resistant strains of bacteria even in the food we eat. They are the ones that survive, reproduce, and grow stronger.

Antibiotic-resistant bacteria can spread into the air from confined-animal feeding operations that use antibiotics to compensate for the overcrowded conditions in their pens, affecting nearby residents. The resistant bacteria in animals' manure can wash downstream and enter waterways where people swim and play. Scientists have even found in the sand on Florida beaches resistant bacteria brought there by seagulls.

The FDA recently announced new regulations to urge drug companies and agribusinesses to phase out the use of certain antibiotics in livestock and poultry, but the regulations are voluntary. And according to Ostfeld, this will definitely not end antibiotic resistance. There are large numbers of antibiotics used for livestock that will not be regulated, and so microbes will continue to evolve antibiotic resistance.

But the resistance issues generated by farm animals are not our only worry. The Cary Institute aquatic ecologist Emma J. Rosi-Marshall has studied how antimicrobial chemicals used in personal-care products leak into the environment. Rosi-Marshall claims that putting antibiotics into toothpaste and hand cleaners serves no health purpose—they're no better than antibiotic-free toothpaste or soap and water—yet they increase antibiotic resistance in the environment.

Common afflictions like gonorrhea have developed resistance to many common antibiotics, including penicillin and tetracycline. Gonorrhea is transmitted sexually between humans. The World Health Organization reports that the disease is becoming a major health challenge in Australia, France, Japan, Norway, Sweden, and the UK due to antibiotic resistance that developed in the late 1990s and early 2000s. Left untreated, gonorrhea can cause painful infections of the reproductive organs, infertility, an increased risk of catching HIV, stillbirths, spontaneous abortions, and blindness in newborns.

Another ailment currently resurging is tuberculosis (TB), a potentially fatal lung disease that has also grown resistant to antibiotics. The bacteria that cause tuberculosis are spread from person to person through tiny droplets released into the air via coughs and sneezes, though you are most likely to get the disease from someone you live with. It was once rare in developed countries, but the number of TB cases has increased worldwide since the 1980s. Part of the problem was caused by the emergence of HIV, the virus that causes AIDS. HIV weakens a person's immune system so it can't fight TB germs.

People who have tuberculosis often must take a variety of medications for long periods to get rid of the infection and deal with drug resistance. Various strains of tuberculosis have been found resistant to medications generally used to treat the disease. Multidrug-resistant tuberculosis rampages through the Russian prison system, where prisoners easily catch the disease and spread it to other inmates. The TB bacterium has developed immunity to many drugs, and has begun to proliferate among homeless people and AIDS patients.

The effects of drug resistance are serious and global. An estimated 630,000 people are presently ill with multidrug-resistant tuberculosis. Some 88 million people are infected with gonorrhea, which is also multidrug-resistant. There are 448 million new cases of curable sexually transmitted diseases (STDs)—including syphilis, chlamydia, and trichomoniasis—every year, and health authorities are watching those diseases for the development of resistant strains.

* * * *

Should drug resistance and a host of new diseases brought on by the elimination of species concern us? How might a major pandemic occur? The influenza epidemic of 1918–19 killed 50 million. The Hong Kong flu of 1968–69 took about one million. The AIDS epidemic has taken some 30 million people so far. It is still a virulent killer in Africa, where the chief victims are now heterosexuals. WHO reports that malaria caused 627,000 deaths in 2012. Right now tuberculosis is making a big comeback.

Michael Greger looks at bird flu as earth's next big catastrophe. Over the better part of the last two decades a killer strain of avian influenza has devastated birds in Asia, Europe, the Middle East, and Africa. It kills more than half of all its avian victims, and some strains kill even more. And it's a virus. It can spread through coughing or sneezing, through the air, just as H1N1 or any of our common viruses can.

In rare instances where bird flu has spread from poultry to people, it's been one of the deadliest viruses ever described. About 600 people have been infected with bird flu and 350 have died, about 60 percent. "But what if the virus were to mutate into easy human-to-human transmissibility?" asked Greger in a televised interview with Thom Hartmann about his book *Bird Flu: A Virus of Our Own Hatching*. "It would be like crossing one of the most deadly diseases, Ebola, with the most contagious disease ever known, influenza."

In 1900, the leading causes of death were tuberculosis, pneumonia, and enteritis. Today, more than a century later, the chief causes of death are heart disease, cancer, and stroke. These chronic diseases have overtaken infectious diseases as our number one killers. This is not a bad thing, since chronic diseases generally affect older populations. Thus the abatement of infectious disease in just the last century has increased average life spans by thirty years or more. The decrease in the role of infectious killers is largely due to inoculations and antibiotics. Some of the greatest recipients of these benefits have been the young, who are disproportionately affected by infectious disease.

But the balance is changing. Recently, Dr. Margaret Chan, director general of the World Health Organization, addressed a group of experts gathered in Geneva, Switzerland, to tackle antibiotic resistance. "Some microorganisms are resistant to nearly everything we can offer to save the lives of infected patients," Chan said in a speech to the convention. "And few new antimicrobials are in the R&D pipeline. Medicines lost because of microbial resistance are not being replaced. We are moving towards a post-antibiotic era where common infections will once again kill. If we lose our most effective antimicrobials [antibiotics, antifungals, antivirals, and antiparasitics], we lose modern medicine as we know it."

Disease is not likely to take man out, any more than the plague, World World II, or AIDS has. But if you take new diseases, couple them with antibiotic resistance, add some rising populations, and mix in a lack of food and proper nutrition, then we might have the recipe for our own extinction.

7

WARNING SIGN III: SQUID AND SPERM WHALES

UNLIKE THE EMERGING THREAT of new diseases and the resistance to antibiotics, one doesn't have to wait to see how man's interference is changing the marine environment. Many of those changes are already here. One shining example is the Gulf of California between mainland Mexico and the Baja California peninsula, what was once lovingly referred to as the "Baja Fish Trap" for its abundance of marine life. Overfishing, acidification, and warming waters have altered the ecology of these famous marine waters. The marlin, swordfish, and sharks that anglers once came here for have dramatically dwindled and a new ecology made up of Humboldt squid and sperm whales has taken over.

It is still a pristine environment. A drive south of the US border down Mexico's Highway 1 takes you past volcanoes, mountains, and sculpted red rock through a series of valleys populated with whiplike boojum trees and giant cardon cacti. About five hundred miles south of the border it summits the coastal mountains and descends rapidly onto the Gulf of California just above the historic French mining town of Santa Rosalía. The Gulf of California, in Mexico, was created six to ten million years ago when Baja California began to separate from mainland Mexico, producing the geologically diverse peninsula and the biologically diverse waters of the Gulf.

On a recent visit, the moist evening breeze brought in the briny smell of marine life to cool the town of Santa Rosalía as the fishermen headed toward the dock and the boats for the nighttime catch. Biologist William Gilly, from Stanford University's Hopkins Marine Station—a big, friendly academic with lots of interesting stories—and his group of student researchers joined the fishermen as they motored out to sea. It was September on the Baja Peninsula, where open-ocean schools of tuna, swordfish, and sharks were once an annual gift of the Gulf, but have diminished in recent years.

Now Santa Rosalía fishermen pursue Humboldt squid (also known as jumbo squid), which appear to have replaced many of the finfish in the Gulf of California. They still fish as before, only they go out in the late evening, not at dawn. At sunset, I watched the local fishermen join the parade of pangas, twenty-two-foot open skiffs with outboard motors that departed from the sandy shores. The Gulf waters turned from blue to black as the boats lined up about a mile offshore, their colored lights glistening in the evening shadows. The fishermen used hand lines baited with fluorescent jigs to catch the squid.

These boats represent a growing group of local small-scale fishermen who, but for their outboard motors, rely little on the hardware of the modern commercial fishing industry. Instead, they fish the waters off the Baja Peninsula from unregulated camps that line the shore using primitive gear. Over the last decade the Mexican Humboldt squid fishery has caught between 50,000 and 200,000 tons of squid annually, mostly from the Gulf of California, and sold it predominantly to markets in Korea and China.

The Humboldt squid was named for the Humboldt Current, an ocean current that flows north along the west coast of South America from the southern tip of Chile to northern Peru. It was thought that the Humboldt squid in Baja originated in Pacific waters off South America, though when, exactly, they arrived off Baja is a mystery.

There have been few historical sightings of the squid in marine records farther north than the Galápagos Islands off South America.

Humboldt squid (*Docidicus gigas*) have not only invaded the waters of the Gulf of California, they have expanded their domain northward along the Pacific coast as far as Alaska and westward along the equator toward the Hawaiian Islands.

Squid here seem to have filled a niche left vacant when finfish such as tuna, sharks, marlin, and swordfish began to disappear in the late twentieth century. Squid have a much shorter life span than other fish, rarely living over a year and a half. And they are highly productive, meaning they can bounce back from fishing pressure much faster than finfish, which are not as productive. But Gilly thought this factor was less important than the ability of squid to cope with the spread of low-oxygen waters, a new problem on the horizon that may be giving the squid their ticket to expand.

The increase in the biomass of Humboldt squid in the Gulf of California is promoted by the development of low-oxygen zones in the water, a result of climate change and possibly decreased ocean circulation. These zones are different from the dead zones created by agricultural runoff, but the two could act in tandem to worsen the effects. Low-oxygen waters support fewer species but can support high quantities of those few species that are tolerant of it. Again, we are seeing the live-fast-die-young generation: a few species that are able to survive a toxic environment, which then take over the world—or the ocean in this case.

Santa Rosalía developed as a copper mining town in the late 1800s, and it was prosperous until the ore ran out in the 1920s. Still there are touches of prosperity from its mining days. Gustave Eiffel, of Eiffel Tower fame, built the church in the town center in France and then shipped it to this Baja town, where it was reassembled in 1897, an indication of the wealth mining generates. Still the town has none of the lights, bars, or tourist trappings you might find in Puerto Vallarta or Acapulco farther south.

The Santa Rosalía copper mine has recently reemerged as newer techniques have made the mining of old ore deposits viable. Gilly wonders what the long-term effects will be as the mine gears up for another run. Only, the proportions are much larger now than in the late 1900s, as miners will be using huge equipment to extract lower amounts of copper from already-mined soils.

Gilly has developed a program for monitoring intertidal shellfish communities, both near the new mine and in a more protected area about twenty miles north of town. "If the mine begins to disturb the marine environment off Santa Rosalía, the monitoring plan is designed to detect it. We're lucky to be able to commence monitoring before major production commences," said Gilly. He's working with students from a local technology school that was established here in recent years.

Still his biggest concern is the changing face of oxygen in the deep ocean, here and in the oceans around the world. Gilly referred me to a paper by Lothar Stramma, a physical oceanographer at Kiel University in Germany, who led a study in 2008 that analyzed oxygen content at six different spots in the deep waters of the Pacific, Atlantic, and Indian Oceans. That study found significant increase in low-oxygen water in most spots, and these areas, known as oxygen minimum zones, were below the livable threshold of many marine animals. These low-oxygen zones are a natural phenomenon of the eastern Pacific Ocean and occur in the upper layers of the water, but they are expanding in all directions worldwide. Scientists link this change to global warming.

The oxygen minimum zone restricts the depth to which tropical open-ocean fishes, such as marlin, sailfish, and tuna, can go by compressing their habitat into a narrow surface layer, where they are more easily fished out. In general, the Pacific has lower oxygen minimum zones than the Atlantic. German oceanographer Stramma said that the lowest oxygen value in the Atlantic found in the 2008 study was 40 percent saturation (surface is 100 percent), whereas in the Pacific there were oxygen minimum zones that reached almost zero percent.

This has serious consequences for marine organisms. According to Gilly, at 10 percent dissolved oxygen content in the water, microorganisms can no longer utilize oxygen and start metabolizing nitrogen compounds, releasing nitrates, which are strong greenhouse gases. "At zero percent, microorganisms start metabolizing sulfate ion compounds and releasing hydrogen sulfide, and that can be lethal," said Gilly. During the Permian extinction the oceans went stagnant in places, caused by a loss of ocean currents. Douglas Erwin at the Smithsonian thinks that the emergence of this chemical compound into the atmosphere may have been one of the dominant killing forces at the time.

Humboldt squid feed on lantern fish in the Gulf of California but may prefer hake in Chile and Peru as well as off Northern California. "Hake" is a term that includes any of several large marine fishes of the cod family. South American authorities struggle with problems in their hake fishery, which is squeezed between overfishing and oxygen-starved waters. Northern California's hake fishery has not been affected by oxygen-starved waters, though bottom-dwelling creatures have.

Off the Oregon and California coasts, the oxygen minimum layer is rising up and moving nearer the shore. "It's intersecting the continental shelf and moving rapidly inland like a river breaching its levees," said Gilly. "And there are a lot of things that live at the bottom that can't swim away."

The presence of large numbers of Humboldt squid off the Pacific Northwest has impacted the valuable hake fishery there. For example, in 2009 there were so many squid present in the areas of hake schools that sonar estimates taken of hake numbers could not be used to set national quotas for the US and Canadian hake fisheries.

Few predators catch squid at these depths. Gilled finfish like tuna and shark can dive to the upper limits of the oxygen minimum zone and feed on squid there, but few can go into the zone and stay there for a significant length of time. Scientists at Stanford University have tracked great white sharks, which migrate annually toward Hawaii, and have found that large numbers of these animals stop en route at a mid-ocean area called the "White Shark Café," where they repeatedly engage in dives above the oxygen minimum zones. Whether they are mating or feeding is not yet known, but Gilly thinks they could be diving for Humboldt squid or the purple-back flying squid that may also inhabit the area.

Fertilizer runoff from the mainland shores in the northeastern part of the Gulf may be enhancing the low oxygen effects here. Such runoff has created dead zones at the mouth of the Mississippi River in the US; the mouth of the Yangtze River in China; within the Black Sea Basin in eastern Europe; in the Skagerrak, the strait that separates Norway and Sweden from Denmark; and in the Cariaco Basin, near the coast of Venezuela. There are more than 150 such dead zones around the world.

The difference between dead zones and low-oxygen zones is that the latter involve an oxygen deficiency in the specific layer of water that forms beneath the maximum depth of daytime surface light in coastal and mid-ocean environments. Scientists measuring that layer of water, between 650 and 3,000 feet (200 and 700 meters), have found a measurable decrease in oxygen and an expansion of the vertical and horizontal limits of the layer over the last fifty years.

This maximum depth of daylight surface light is also known as the deep-scattering layer, a name given to it by twentieth-century naval captains who found that sonar gave a false seafloor echo as it bounced off this zone because of the high density of marine life present. Plankton and zooplankton congregate in the deep-scattering layer primarily to avoid visual predators, and their feeding habits use up dissolved oxygen in the water, creating the oxygen minimum zones.

Few marine creatures have adapted to the oxygen minimum zones. But Humboldt squid are one of these low-oxygen-tolerant wonders. When they enter the zone, their metabolism slows and they consume less than 20 percent of the oxygen they need at the surface. Specialized gills allow them to scavenge oxygen from the water more efficiently. Their hearts don't race wildly as they chase down their prey, since their prey are slowed down by the lack of oxygen as much as the squid are. "It's not like a lion chasing after a gazelle," says Gilly. "They catch fish with little effort."

What are known as "common market squid," a smaller but important part of the California fishery, probably find such zones lethal. Gilly, who has studied both common market squid and Humboldt squid for decades, believes that increasing loss of oxygen in the seas will lead to the expansion of Humboldt squid from this point forward. This is bad for finfish, as the larger fish—already crowded into shallower oxygen-rich zones—will become more vulnerable to commercial fishing. Such a situation is happening now off the coast of Peru and Chile around the Humboldt Current, one of the richest fisheries on earth, where catches are high but the sustainability of these catch rates is in doubt.

Climate change is the chief suspect in this developing tragedy. Warmer ocean waters hold less oxygen, and a warmer climate generates less wind to oxygenate surface waters. The result is a more stratified ocean with a surface layer of warm water riding on cooler, denser water, which impedes the mixing of oxygen. In addition, shrinking ice at the poles may be slowing deep-ocean circulation, which brings oxygenated waters to the deep waters of the Pacific and Atlantic Oceans.

During that Permian extinction 250 million years ago, increased atmospheric CO_2 warmed the planet, which stripped the ocean of its oxygen and wiped out more than 90 percent of the creatures in the sea. Oxygen deprivation was a major source of extinction during the Cretaceous extinction as well.

Bigeye tuna, swordfish, and sharks can dive to the top of the oxygen minimum zone, but few finfish can go into it for any length of time. Sperm whales, elephant seals, and some sea turtles are among the best penetrators of this zone, but it takes serious adaptations to withstand the pressure and the lack of oxygen. For the few that can, the upper boundary of the oxygen minimum zone is a hidden treasure where life abounds.

FOLLOWING STEINBECK

To show the extent of change that has occurred over the last half century, Gilly likes to refer to descriptions by the author John Steinbeck and the marine biologist Ed Ricketts, who in 1940 took a trip around the Baja Peninsula into the Gulf of California surveying the marine life. Steinbeck wrote a book about the journey, called *The Log from the Sea of Cortez*—the Sea of Cortez being the more traditional, more romantic name for the Gulf—describing his trip with Ricketts and a crew of fishermen from Monterey, California. Steinbeck had featured Ricketts, who made a living at his lab on Cannery Row by preserving specimens of marine life and selling them to schools for use in biology laboratories, in two of his novels, *Cannery Row* and *Sweet Thursday*.

The purpose of the Steinbeck/Ricketts expedition in 1940 was to collect samples in the tide pools along the shores of the Gulf of California over a six-week tour. The group left Cannery Row in Monterey at a time when Hitler was invading Denmark and moving up toward Norway and "there was no telling when the invasion of England might begin," wrote Steinbeck. But they put the world's drama in their rearview mirror, and boarded the *Western Flyer*, a chartered sardine boat, heading for Baja California, Mexico.

Three days later, they eyed the lighthouse at Cabo San Lucas, at the southern tip of the peninsula, and at about 10 p.m. they rounded the cape and entered the dark harbor. Except for the lighthouse, there were no lights in the harbor. Today, Cabo San Lucas is a full-blown mega-resort, with lights that stay on all night. Then it was a sleepy little village where it took Steinbeck and Ricketts all day to find the authorities in order to get their visas stamped.

The first Mexican town Steinbeck described at length in *The Log from the Sea of Cortez* was La Paz, a large port around the southern tip of Baja coming from the Pacific. I visited La Paz last summer and witnessed the various efforts being made to compensate local fishermen for the reduced catches they and their hungry families are encountering.

Frank Hurd is the science director of Olazul, a group of American and Mexican scientists and innovators working with local fishing communities to develop sustainable systems of aquaculture as an alternative to depleting overfished stocks. Hurd invited me to see his version of an offshore, semimobile aquaculture pen. One morning before dawn we drove out from the city to a fish camp on the northern shores of La Paz where Hurd and his associates had been testing a spherical pen, 277 cubic yards (212 cubic meters) in volume, about three miles offshore. Hurd said Gulf currents could flush out the wastes and bring in nutrients and oxygen for the shrimp he was testing. The structure was made from recycled and reinforced polyethylene timbers wrapped in coated steel mesh netting "built to withstand the occasional hurricane that rolls up the Mexican shoreline during the summer and early fall," said Hurd.

In his book, Steinbeck had described on their Sea of Cortez journey how they trolled a couple of lines off the back of their boat and were pretty much able to keep themselves in finfish such as yellowfin tuna, skipjack, Mexican sierra, red snapper, and barracuda the whole trip. Hurd said that the local fishermen in La Paz described similar catches in the old days but today try to make a living selling trigger fish, sand bass, bonito, mackerel, and other species that were considered trash fish back in Steinbeck's time.

The Sea of Cortez that Steinbeck investigated over seventy years ago is not the same body of water that Gilly and his crew motored to in 2004. In his log, Steinbeck described marlin and swordfish frequently leaping out of the ocean into the air and dancing across the surface of the water. The scientist described seeing only a couple of small squid on his entire six-week excursion in the Gulf of California. And there was nothing that resembled a Humboldt squid.

Gilly also spent some time looking through historical records for sightings of Humboldt squid. There were isolated reports in the scientific literature going back to 1938, but no reports of large numbers until commercial fishing for them commenced in the late 1970s. He queried a number of old fishermen in the Gulf, and none of them remembered sighting the squid before that. Humboldt squid were absent in the natural history of the Gulf written by early Jesuit missionaries. James Colnett, an officer of the British Royal Navy, saw no Humboldt squid in the area south of Cabo San Lucas in 1793–94, though he described squid "of four or five feet in length" at the surface off the Galápagos Islands. "But that's a far way off," said Gilly. The squid must have migrated to the Gulf of California since Colnett's time, but details of the move are lacking.

Humboldt squid appear to have evolved in the southeastern Pacific where El Niño events warm the surface waters of the ocean every four to twelve years, creating unusual global weather patterns. Changes in the Humboldt squid fishery mirror changes in El Niño–driven weather. Though Gilly and his associates measured high concentrations of squid in the central Gulf in 2012, they had moved away from the shore and their sizes had decreased. Gilly thought an earlier El Niño event in 2009–2010 led to the animals' accelerated sexual maturity, what he called an even more radical live-fast-die-young life strategy in the face of an uncertain future.

Humboldt squid have two tentacles that can reach out and grab prey and eight arms to envelope them. The squid can attain eight feet in total length (mantle plus tentacles). They use their tentacles and arms to subdue prey and their razor-sharp parrotlike beaks to tear them apart. They are some of the fiercest of the cephalopods, a group of animals that includes squid, cuttlefish, and octopus.

Humboldt squid are also famously cannibalistic. Unai Markaida, a marine biologist at El Colegio de la Frontera Sur in Campeche, Mexico, studied prey items of 533 Humboldt squid and found evidence of other Humboldt squid in 26 percent of their stomachs. Fishermen who pursue the Humboldt squid tell scientists that once squid are hooked, other squid start attacking and eating the fishermen's catch.

The fishermen have to pull their catches in fast to avoid these voracious attacks.

The Humboldt squid is particularly fast and propels itself through the ocean as if by jet engine. It draws water into its mantle and then ejects it through a spout like a rocket. All squid have the ability to change color quickly, some imitating patterns, even textures of sandy bottoms or rocky reefs. Humboldt squid lack this patterning capacity but are able to switch back and forth from maroon to ivory, pulsing like a strobe. The capability to communicate through color change is quite profound for a creature that is related to the snail. According to Gilly, "There's jitter [vibration], variation, and change in the frequency between two squid. It's highly unlikely this isn't some kind of communication."

Up to four million Humboldt squid hang out in the Sea of Cortez near Santa Rosalía at about one thousand feet (three hundred meters) during daylight near the shelf where the bottom starts falling off sharply, but move up at night when the deep-scattering layer moves up as well. It's then that the fishermen initiate their attack. Hauling up a squid that can weigh up to a hundred pounds by hand lines is a rough job at night, particularly when the average price for cleaned squid is less than ten cents a pound (0.5 kilos).

One winter day, I caught up with Bill Gilly at the Hopkins Marine Lab on Cannery Row, next to the Monterey Bay Aquarium. People claim there are differences but also a lot of similarities between Gilly and Ed Ricketts, who accompanied Steinbeck on his journey to the Sea of Cortez.

Ricketts's lab on Cannery Row was a hangout for authors, illustrious locals, and street people. Gilly's lab is more of a gathering place for assorted Stanford students. Ricketts, according to Steinbeck, lived across from the local house of prostitution but never visited the house after dark unless he'd run out of beer and the stores were closed. Gilly lives next to the Monterey Bay Aquarium and goes there frequently.

In Steinbeck's eulogy he said Ricketts "loved to drink just about anything." Gilly would admit only to enjoying the occasional beer. Their greatest similarities are that Gilly, like Ricketts, is a biologist who loves Monterey, Baja, and the Gulf of California and likes to laugh.

When Gilly announced his plan to retrace Steinbeck and Ricketts's 1940 voyage through the waters and intertidal zones of the Gulf of California, he received a surprising call of support when the owner of North Coast Brewing Company called and offered his services. "You know those guys drank a lot of beer on that trip," he said. "And I'm the man that can help you with that." Gilly reacted with a smile.

Gilly and his team arrived at the boat on their date of departure and found two shrink-wrapped pallets of beer with a sign on it that said FOR DOCTOR GILLY. Inside the shrink-wrap were seventy-two cases. "It was the most beer I've ever seen outside of a Princeton reunion I once attended," laughed Gilly when he told me the tale.

In the end, Gilly and his crew drank only about 1,242 beers. Steinbeck and his crew drank 2,160 beers. "And they did that with a smaller crew and a shorter trip," said Gilly in awe.

There were other differences in the two expeditions that were not as lighthearted. At the various intertidal zones that Steinbeck and Ricketts visited, the author repeatedly used expressions like spiny-skinned starfish in "great numbers," and "knots" of brittle stars, but Gilly's team did not observe a great number of either at any of the tide pools they witnessed.

Steinbeck and Ricketts encountered "huge" conches and whelks (large sea snails and their shells) at several sites and a great number of large *Turbo* snails (shaped like a turbine). Gilly's crew found only small living specimens of conches and *Turbo* at just a few sites, and dead whelk shells at one. In 1936, William Beebe, an American naturalist, explorer, and marine biologist, found a beach just north of Bahía Concepción—about midway up the eastern shore of the Baja—that was what he called "a conchologist's paradise," with shells "of amazing size and a host of species." Tellingly, Gilly's crew found a dramatic decline in all these species.

One of the greatest and most disturbing changes in the Gulf is in the "pelagic," or open-ocean, predatory finfish that inhabit the upper portion of the water column and aren't associated with the shore or the bottom. Although Gilly and company traveled at the same time of year, using the same type of boat, and for about the same duration as the Steinbeck adventure, they witnessed a greatly changed community of open-ocean fish.

Steinbeck and Ricketts wrote, "We could see the splashing of great schools of tuna in the distance, where they beat the water to spray." The pair also saw marlin, sailfish, and swordfish, but Gilly's team sighted few of these.

Gilly's team did catch sierra mackerel and yellowtail, but neither fish was of the same size or in the same numbers as Steinbeck and Ricketts had reported.

Steinbeck and Ricketts got a look into the future, though they didn't realize it at the time, when they boarded a shrimp trawler off Guaymas on the mainland side of the Gulf and witnessed the unintended consequences of the fish species that came up with each net. Though the fishermen tried to separate the shrimp from the rest of the catch and tossed back the unwanted fish, most of these died belly-up in the water. Today shrimp trawling is recognized as the single most ecologically damaging activity in the Gulf.

Sharks, particularly the enormous schools of hammerheads that once circled the sunken islands, or seamounts, in the middle of the Gulf, have declined in size and number. The same is true with manta rays: they have been replaced with smaller mobula rays than the ones seen by Steinbeck and Ricketts. Steinbeck wrote of the attempted landing of a number of huge manta rays, but the rays always broke the line, even when it was three inches thick. And Steinbeck and Ricketts noted several other species of squid but not Humboldt.

Though Gilly never saw the abundance of marine life that Steinbeck had witnessed, he found his own vision when he got to San Pedro Mártir Island, an area known for hosting many sperm whales. Since sperm whales eat Humboldt squid, Gilly figured there must be a lot of squid, and this was the reason for taking this detour, which had not been a part of Steinbeck's journey. He was looking for baby Humboldt squid, something nobody had ever found in the Gulf. Satellite data had told him there was an intense tidal upwelling event (tides bringing the rich waters of the deep up toward the sea surface), and he guessed that the forward edge of this rich marine zone might harbor tiny squid larvae. On his second net tow, he found two baby squid a quarter of an inch long. But there was more.

It was a place where "all the life was, plankton, fish, squid, and whales," reported Gilly. The biologist and his team were greeted by a nonstop squid review, with Humboldt squid darting in toward the boat and flashing their underbellies in attempts to lure small schooling fish near the surface. The show continued until after midnight.

They didn't set anchor, since the sea was more than 3,300 feet (1,000 meters) deep, so they simply drifted all night. At one point, large sperm whales were lounging at the surface with fins exposed, some in pairs, showing their flukes before diving. Gilly had never seen a sperm whale before, yet he knew they were hanging out because Humboldt squid, their favorite food, were there in abundance.

Said Gilly, "We had come to the Sea of Cortez to discover how things might have changed since 1940—and here on the open water was the most dramatic ecological change witnessed during the entire trip, the apparent arrival of two major predators far offshore from the rocky reefs that were scoured by Ricketts and Steinbeck. This was a profound and qualitative change—an ecological regime shift."

It was the apex of a new evolution, one made up of squid and sperm whales, which had replaced the vision of tuna, marlin, sailfish, sharks, and other finfish that Steinbeck and Ricketts had seen only seventy years earlier.

The Gulf of California was not the only place experiencing ecological change due to low-oxygen waters. Beginning in 2002, low-oxygen water from off the North Pacific shore had slipped up over the continental shelf and moved inshore, killing off bottom-dwelling marine creatures off the coast of California, Oregon, and southern Washington. Gilly and others have been watching these findings, too. These low-oxygen events normally arrived in the late-summer months.

In 2006, Pacific waters off Oregon went into an anoxic (no oxygen) condition, killing off many organisms. Submersible vehicles put into the water recorded dead fish strewn across the bottom. Surveys revealed near complete mortality of bottom-dwelling creatures. Continental shelf waters off the Pacific Northwest are from twenty to fifty miles across. They lie beneath the California Current, which is one of the richest marine ecosystems in the world, though this system will be in jeopardy if low-oxygen events grow in size and frequency.

Waters flowing south along the shore tend to bank clockwise in the northern hemisphere and counterclockwise in the southern hemisphere, an effect caused by the rotation of the earth on its axis. Off the US Pacific coast, prevailing northwest winds push surface water away from the coast, and cold, nutrient-rich waters at depth are pulled up to replace it. This greatly increases the productivity of coastal marine environments.

Though many fish stocks are down off the California coast, marine mammals are doing well. Part of this may be the result of their adapting to consume squid and other creatures in the deeper sea. To get to these depths, whales, dolphins, seals, and sea lions all had to pass through a unique event in evolution. In prehistoric times, ancestors of these mammals came out of the sea as fish, lost their gills, and evolved lungs to breathe air. But they returned to the sea when competition from land animals increased, and they had to learn to survive underwater again, only this time breathing air. They currently use a host of neat breath-holding tricks, since the deep ocean is not a friendly place for air breathers. Gilly says that studying sperm whales is difficult, so scientists study other diving marine animals as proxies for whales.

DEEP DIVERS

Back in the 1960s, scientists generally thought animals might dive to 325 to 650 feet (100 to 200 meters), but researchers at the Scripps Institution of Oceanography in San Diego, California, recorded a Weddell seal in McMurdo Sound off Antarctica that dived to 1,970 feet (600 meters).

Since then, emperor penguins, leatherback sea turtles, northern elephant seals, bottlenose whales, and sperm whales have met and surpassed that record. In a world where the oceans may no longer hold enough oxygen for gill-breathing fish, breath holders might still have a chance at survival.

Elephant seals are a great example of an adaptable breath-holding mammal. And since they come out of the water only twice a year for extended stays, they are a lot easier to follow by attaching tags and transmitters while on land in order to record dives. Northern elephant seals have recovered from near extinction and number nearly one hundred thousand animals in the North Pacific. They utilize a unique set of evolutionary adaptations on their deep dives. Their heart rates at the surface are about 120 beats per minute, but while diving they can reduce that to 30 to 35 beats. They have even been recorded as low as 2 beats per minute—the edge of cardiac arrest in a human. Unlike man, most of the oxygen in diving elephant seals is stored in myoglobin in the muscle and hemoglobin in the blood instead of the lungs. They have higher concentrations of hemoglobin in the blood and larger blood volumes than most animals.

Elephant seals have streamlined bodies that glide through the water as if they were traveling on a layer of ball bearings. In a paper in *Nature* in 2011, biologists at the University of California, Santa Cruz, reported that one elephant seal dove to 5,765 feet, then a record for the species. That's the equivalent of more than three Empire State Buildings stacked on top of one another, with the seal plummeting from the top of the uppermost building to the basement of the bottom building before coming back to the surface, a distance of over two miles. Aside from the two or three months of the year that they come out on land to mate or to molt, they are mostly underwater, not really a diving animal but more of an animal that occasionally surfaces.

All the elephant seal's air passages, including the lungs, collapse flat and become airless between 350 and 700 feet (100 and 200 meters). With no air in these spaces, there can be no exchange of gas (particularly nitrogen), and thus elephant seals avoid the blood chemistry imbalances such as the bends (nitrogen bubbles) and rapture of the deep (nitrogen poisoning) that plague human divers.

On these dives, the elephant seal's face looks like a prune. Researchers like to paint up Styrofoam mannequins, put lipstick on them, color their eyes, and send them down 300 feet just for kicks. They come back looking like shrunken heads. But the dives are worth it to the elephant seals, said University of California biologist Burney Le Boeuf: "The deep-scattering layer, the top of the oxygen minimum zone, is where most of the biomass in the ocean is concentrated. These animals are diving to the center of the richest part."

But it's dark down there. Cameras attached to these animals come back with images of a black screen. Some fish are bioluminescent, like lantern fish, a favorite of Humboldt squid in the Sea of Cortez. Whales and seals may swim below their prey and in daylight hours look back up at their silhouettes. The animals are well adapted for this territory. The enormous eyes of the elephant seal help it see in the dark. Whales may do one better than elephant seals, using natural sonar systems to locate their prey. The nose of the sperm whale constitutes a quarter to a third of its total weight and may contain the most powerful sonar system in the natural world.

Deep divers have a virtual monopoly on their prey at those depths, and they also avoid two deadly predators that spend most of their time at the surface: great white sharks and killer whales. For the most part, elephant seals are attacked when getting in and out of the waters on the islands they visit twice a year—for a month or two in winter to breed and for a month or less in summer to molt. They spend the rest of the year in the water on long northern migrations of up to thirteen thousand miles. They dive almost continuously on these trips, each dive lasting twenty minutes or more, after which they spend two or three minutes at the surface, taking in oxygen and letting out CO_2, before they head back down. These are incredible adaptions for an animal that breathes air.

Without these threats, diving can be almost an autopilot affair. Sperm whales and elephant seals sleep as they dive, closing one eye while half of the brain naps and the other side keeps vigilant, then switching back and forth. Plus, once they get down to those depths, the escape responses of prey are a lot slower, allowing deep divers to wander around as if they were at an all-you-can-eat buffet.

But once again the remarkable Humboldt squid has another adaptation in its bag of tricks. Gilly worked on a study with biologist Julia Stewart and found that Humboldt squid, both off Monterey Bay and in the Gulf of California, sometimes power-dive to depths of up to one mile—right through the oxygen minimum zone—and remain there for long periods of time, sometimes all day, before powering upward again. This trick is possible because the oxygen minimum zone is really a layer, and oxygen starts going up again at depths of more than about 3,500 feet because of deep ocean currents that bring oxygen to deep waters. "These extraordinary dives by Humboldt squid may be escape responses triggered by the presence of groups of foraging marine mammals. The squid simply dive down, hang out for several hours, and then pop back up, hoping to find the predators gone," says Gilly. Only squid seem to navigate these low-oxygen zones so effortlessly. Seals and whales have to come back up for air, but squid can move up and down without it.

Most fish, however, are limited to shallower waters, where they are the target of marine mammals and man, the latter responsible for diminishing fisheries. According to the World Wildlife Fund, the Gulf is the source of nearly 75 percent of Mexico's total annual fish catch, but overfishing (both industrial and artisanal) is contributing to dramatic declines in sharks, rays, and finfish. The global decline in fish catches, combined with rising demand, is leading to a global fishery crisis that threatens the Gulf of California as well as the rest of the world.

Humanity doesn't limit its impacts to fish most commonly found on menus. Exotic sea creatures from turtles to manta rays to marine mammals are being hunted to extinction. Shark numbers, for example, have declined by 80 percent, with one-third of shark species now at risk of extinction. The top marine predator is no longer the shark; it is us.

It has been ten thousand years since most humans lived as hunter-gatherers. Fish are the last wild animals that we hunt in large numbers. And yet we may be the last generation to do so. On average, people eat four times as much fish now as they did in 1950.

In the late 1980s, the photographer George H. H. Huey and I went to the end of Baja to do a story on the shark fishermen there, who complained of fewer and smaller shark catches. When I interviewed a number of marine biologists about this, no one could imagine that these great open-ocean species could diminish. Even Rachel Carson couldn't imagine fish stocks diminishing. But all that has changed.

Humboldt squid could beat the odds against other marine creatures. Their numbers are expanding at sea, while ocean fish populations are contracting. In a relatively short period of time, this squid has learned to adapt to climate change and alterations of oxygen content in the water, conditions that are fatal for many other animals.

Gilly has a lot of respect for this animal: "If someone wanted to design an ocean predator for the future, this would sure be it."

What evolutionary adaptations will we need to survive our future?

Part III

NO-MAN'S-LAND

8

THE END

THE LOSS OF THE diversity of life on earth has implications for man that we appear ready to ignore. Mammals, reptiles, birds, amphibians, and fish possess what are called "ecosystem services," functions they perform that are crucial to the well-being of nature and *Homo sapiens*. Their loss is our loss. Without their survival, ours is in question. It's why some scientists believe man won't survive a mass extinction, because of all the ecosystem services we will lose in such an event.

We've seen how diversity of forest animals can help protect us from disease, but this is not nature's only gift for our survival. Other living things, like plants, insects, and microbes, play vital roles in our lives as well. One of those valuable roles is creating clean water. New York City's drinking water, which is naturally cleansed on its 125-mile journey from the Catskills to the city, is an example. Many of the system's best purifiers lie beneath the forest floor: in the fine roots of the trees filtering the water, and in microorganisms in the soil that break down contaminants. These natural processes in the watershed absorb as much as half of the nitrogen coming into the waterways from auto emissions, fertilizers, and manures. In the wetlands section of the water's travels, cattails and other plants also help filter nutrients as they trap sediment and heavy metals.

New York's system of waterways owes its existence in part to an epidemic of Asiatic cholera, which in 1832 killed nearly one in fifty of the city's inhabitants and prompted more than half the population to leave town. New York City politicians quickly launched the construction of a major drinking water system by damming the east and west branches of the Croton River, forty miles upstate in Westchester and Putnam Counties, and then built aqueducts to channel that water to reservoirs in downtown Manhattan.

But New Yorkers were still thirsty. So the city's Board of Water Supply looked farther out of town to the Catskill Mountains. Today, New York City's source is the Catskill/Delaware Watershed, named after the two rivers that have delivered water to the city for most of the twentieth century. The watershed provides drinking water to nearly ten million people, and for a long time its supply has been kept clean by natural filtration. But in 1986 the US Congress amended the Safe Drinking Water Act, which was originally passed by Congress in 1974 to protect public health by regulating the nation's public drinking water supply. The amendment pressed New York City to build a $6 billion to $8 billion filtration system. Instead the city proposed protecting this valuable watershed by buying land as a buffer and a natural filter while upgrading sewage treatment plants.

But housing development got in the way. Roads and homes started to appear in the Catskill/Delaware Watershed, and New York City politicians procrastinated on their land purchase proposal. To get things going, Robert F. Kennedy Jr., son of the late senator, then the attorney for New York's clean water advocate, Riverkeeper, solicited a real estate agent who estimated that it would cost only $1 billion to buy every acre in the Catskill/Delaware Watershed, several billion dollars less than a filtration system. The real answer, Kennedy told one reporter, was to "stop development. That's what you have to do, but nobody wants to say it."

Kennedy kept pushing the city on Catskill land purchases, taking film crews into one faulty hospital treatment plant, showing how sewage and wastewater were leaking out into the New York system.

The *New York Post* reported that the Croton reservoir had been shut down due to pollution by sewage, but a New York City spokesman countered that it had been shut down by "organic material." The late-night television host David Letterman joked that the story "scared the organic material out of me."

New York City reacted by putting severe restraints on development, new sewage plants, paved surfaces, and farming activities in the watershed, but local residents countered with lawsuits alleging they were being asked to shoulder the cost of New York's drinking water. The battle ended in a compromise in which the city promised to spend $1.5 billion to buy up land and to construct and repair necessary storm drains and sewage systems. The EPA put off the New York City requirement to build a drinking water filtration system for another five years.

Today the city does everything to guarantee the safety of its water, including following urban sales of Pepto-Bismol and Imodium, both dysentery medicines, to help monitor water quality. Inspectors look for outbreaks of disease caused by single-celled parasites such as *Giardia lamblia* and *Cryptosporidium parvum* in the city's water supply. *Giardia* can cause cramps and diarrhea, but just one cyst of *Cryptosporidium* can lead to severe illness or death in people with weakened immune systems.

Right now nature is producing the correct amounts of plants, forests, cattails, earthworms, and soil bacteria to keep these and other illnesses out of New York. But if we keep destroying species, the biological equilibrium of these natural systems won't be there to offer its first line of defense.

How do we destroy species in a watershed? Lots of ways. Invasive pests such as the emerald ash borer, the gypsy moth, and the Asian long-horned beetle threaten Catskill trees. Pollution runoff can overwhelm wetland abilities to trap sediment and heavy metals, and if forests and wetlands go, so do the filtration efforts of the plant roots. Climate change is reducing snowfall in the Northeast, and this exposes the roots of trees to colder temperatures than they would experience under a blanket of snow, and this can lead to diminished watershed trees. And diminished trees mean diminished microbial communities beneath them.

Protecting natural environments for the sake of their ecosystem services isn't just a trendy New York City idea. Boston escaped an order from the EPA to filter its water by enacting a watershed program similar to New York City's that included land purchases, wildlife control, and the regulation of development along tributaries. In Costa Rica, the government charges customers a few cents more on their monthly water bills to pay upstream farmers to preserve and restore the tropical forest. The European Union requires watershed protection of woodlands to ensure the quality and clarity of its water.

In the late 1980s, Perrier water in northeastern France began protecting the Rhine-Meuse watershed for fear that pesticides and fertilizers would compromise the quality of its famous bottled water. In 1990, the water was temporarily pulled off the shelf when it was found to contain the carcinogen benzene, a component of gasoline. Rather than relocate, Perrier spent $9 million to buy six hundred acres around its famous spring. They also entered into long-term agreements with local farmers to use more environmentally friendly practices on four thousand more acres of surrounding land.

Though there is a substantial amount of knowledge about the importance of natural systems to the human economy, the idea hasn't entered the consciousness of public and political minds. Ecosystem services are the processes by which natural ecosystems and the species they contain sustain human life. They bring us seafood, forage, timber, biomass fuels, natural fiber, pharmaceuticals, and more.

Critical services could include the purification of water and air, mitigation of floods and droughts, breakdown of wastes, generation of soil, pollination of crops, control of agricultural pests, dispersal of seeds, protection from the sun, moderation of temperature, winds, and waves, as well as enough aesthetic beauty to lift the human spirit.

That's a lot of important functions. There are legions of ecosystem soldiers contained in some of those goods. One square meter (1.2 square yards) of Denmark pasture, for instance, is populated with approximately 50,000 earthworms, 50,000 insects and mites, and nearly 12 million roundworms. A single gram of soil has about 30,000 protozoans, 50,000 algae, 400,000 fungi, and billions of individual bacteria. These life-forms perform complex natural cycles that are critical to human life.

Without birds and other insect predators, pesticides alone could not control agricultural pests. Without pollinators, plants would not produce food. But many of our ecosystem "soldiers" are in trouble. Nearly twenty thousand species of animals and plants are presently considered at high risk of extinction. A study in *Nature* concluded that if all the species that were considered threatened were lost in this century, and if the rate of extinction continued, we would be on track to lose three-quarters or more of all species within the next century. The International Union for Conservation of Nature has evaluated more than fifty-two thousand animal and plant species for their ability to survive. Their conclusion is that 25 percent of mammal species are threatened, as well as 13 percent of bird species, 41 percent of amphibian species, 28 percent of reptile species, and 28 percent of known fish species.

Yet we are dependent upon these species for our own survival. Ecosystems of multiple species interact with one another and their environments, and those interactions are essential for human life. They represent the genetic diversity of life, providing the raw ingredients for new medicines, new crops, and new livestock.

Forests store more carbon from carbon dioxide if they have a greater variety of tree species. Streams clean up more pollution if they have a greater variety of microbes. Increasing the diversity of fish means there are greater fishery yields. Increasing plant diversity means they can better fend off invasive plants. Natural enemies better control agricultural pests if they are composed of a variety of predators, parasites, and pathogens. And ecosystems with a greater biodiversity can better withstand stress such as higher temperatures.

On the other hand, less diversity means less carbon capture, more polluted streams, fewer fish, more invasive plants, more agricultural pests, and more of the species that do poorly under stress.

There is a cultural aspect to ecosystem services as well. Madhav Gadgil of the Indian Institute of Science, Bangalore, and Kamaljit Bawa of the University of Massachusetts Boston divide the consumers of the world into two categories: there are ecosystem people who include forest dwellers, herders, fishers, and peasants who rely on local ecosystems to fulfill most of their needs; and there are biosphere people who extract ecosystem products over a larger international range for commercial purposes. Their rewards are uneven. Even when ecosystem people extract local products for biosphere people, they often do so for low wages because they don't own the land or the trucks, trains, and airplanes to get the products to commercial markets.

Communities relying on local goods have an incentive to conserve the products so they are there tomorrow. But what happens, says Gadgil and Bawa, is that local people without a controlling interest in the ecosystems nearby aren't as involved or committed to the long-term survival of these ecosystems. Approximately fifty million people in India live in proximity to forested areas and derive the majority of their living from forest products. But they often don't have ownership of the lands or the goods.

According to Gadgil and Bawa, if restoration of the environment is to be paramount in economic decisions, then some of that locally controlled, locally extracted, and locally used philosophy has to rub off on the biosphere people. Just because one enjoys blueberries in the middle of winter doesn't mean that it's a good or healthy idea to buy produce that comes from the other side of the globe. Next time you see food that is shipped more than a hundred miles to your door, think about all the pollutants that are coming out of the back of that truck or the fuselage of that plane to get to where you are in winter. It may be benefiting some corporation, the biosphere people, but it's not beneficial for local economies or local health. And with climate change, local health and local economy are vitally connected with international health and economies.

Training our tongue to enjoy foods that are local and in season is healthier for us all. Says Julia Kornegay, a horticulturalist at North Carolina State University, "Trying to have strawberries and raspberries 365 days a year and expecting them to taste good isn't sensible."

Consider also the ecosystem services we derive from diverse tropical plants through the development of medicines. Fifty percent of all medicines owe their origins to species of either plants or animals. Those include tranquilizers, diuretics, analgesics, antibiotics, and more. Aspirin owes its origination to the willow tree. The contraceptive pill originally comes from the wild yam, which grows in the Mexican forests. The bark of the yew tree in the US Pacific Northwest contains the biological compound for Taxol (paclitaxel), which attacks cancer cells that don't respond to other drugs. Madagascar's rosy periwinkle has fostered two different drugs that have altered the outcome of a child with newly diagnosed leukemia from one chance in ten of remission in the 1960s (before these medicines) to nineteen chances in twenty today (after their discovery).

Anticancer drugs derived from plants save about thirty thousand lives each year, with an economic savings of $370 billion in terms of lives saved, suffering reduced, and work hours maintained. Many recent anticancer drugs have been found in the tropics, but unfortunately this is also where the majority of plant species extinctions have occurred.

Norman Myers at Oxford points out that Eli Lilly, a global pharmaceutical company, exploiter of the rosy periwinkle for two anticancer drugs, has profited with over $100 million a year in sales going back to the 1960s. Madagascar, where the plant was taken, hasn't received any of that. This gives that country little incentive to protect the remaining tropical forests even though they may contain the seeds of discovery for a host of other important pharmaceuticals. *Homo sapiens* evolved from an ancestor who hunted for a living, going to each new area, killing the animals, and using the plants. Though our technology has rapidly expanded, our primary instincts are back in the Stone Age.

The forest itself is part of our treasure chest of natural resources, one that has many ecosystem services to offer, but again one that we fail to appreciate. If there are trees by the road or in our neighborhood, then all is well. But if deforestation occurs off the road, in other states, or other countries, we object less forcefully. Out of sight, out of trees.

A prime example of man's selective values can be found in the forests of Central America. I met Dalia Amor Conde, an assistant professor at the Max-Planck Odense Center, in Odense, Denmark, at Flores, a city in the northern department (province) of Petén, Guatemala. The city is located on an island in Lake Petén Itzá, just outside Tikal National Park, famous for its Mayan ruins and its wildlife. Conde was born in Mexico and got her PhD at Duke University, where she began studying jaguar movements in the tropical forest of Central America to determine their habitat and how roads and other infrastructure planned for the region might affect them.

Her goal is to save enough contiguous land to allow jaguars to migrate between isolated populations, keeping the gene pool of the animal mixed and vital. In the process, this encourages the preservation and vitality of a host of plants and animals that reside in the same ecosystem. It is known as the umbrella effect. By saving this charismatic species, Conde hopes to also save the multitude of animals, plants, and birds that reside under the jaguar's umbrella.

On a misty tropical morning, she took me to a local zoo in the middle of Lake Petén Itzá. The zoo had a spotted jaguar whose head and muscular body looked quite regal. We both squatted down to get an eye-level look at the animal, though the jaguar ignored our presence while it paced around its enclosure. This cat is the third-largest feline in the world after the tiger and the lion, and the largest in the western hemisphere. Conde had been on a number of expeditions in the tropical forest whose purpose was to capture jaguars and then release them into the wild with radio collars so biologists could track their movements.

She described to me one hunt in the rain forest surrounding the Mayan ruins of Calakmul in the Yucatán, where birds filled the air with their calls and howler monkeys roared from the treetops. She'd accompanied a caravan of vehicles filled with four biologists, two trackers, five dogs, and a veterinarian to check bait stations along a dirt road through Calakmul National Park. The bait consisted of large chunks of sterilized goat meat spiked with enough drugs to slow the animals down. They were placed every mile at seven spots along a dirt road.

The tracker was Tony Rivera, a former jaguar hunter and now director of EcoSafaris. He got out of his car and announced that the big cat had taken the bait, and the frenzy of dogs in the back of the truck told him the animal was near. Though the group had been up since 3 a.m., everyone suddenly came alive, piled out of the cars, and readied for the hunt.

Rivera let the dogs go, and they took off into the jungle—the biologists doing their best to keep up. As the sound of the dogs' barking changed, Rivera quickened his pace, approaching a tree in which the jaguar had taken refuge. The dogs were pulled back. Rivera raised his rifle, took aim, and fired a tranquilizer dart into the animal's side. Soon the drug took effect, and the biologists were assessing the cat's health and attaching a radio collar to its neck to follow the animal's movements.

Conde found the process transformative. "The first time I looked into the eyes of a jaguar changed my life forever," said Conde.

Conde works with the Mexican NGO Jaguar Conservancy and the

National Autonomous University of Mexico to save the Mesoamerican forest, which runs from Panama to Mexico, the largest remnant of rain forest outside of the Amazon in the western hemisphere. And they are doing this by preserving the jaguar, an animal with a lot of cachet in Latin America.

She was trying to pinpoint specific areas of forest with high populations of jaguars, to make sure they were connected to areas where populations were low. On the boat ride back from the island zoo, Conde said, "With so little of the forest left, the connectivity between the patches is critical. We have isolated populations of jaguars in a sea of human land use."

The habitat of the jaguar, which once ranged from the southern boundaries of the US all the way to Brazil, has shrunk by 80 percent in the last one hundred years. Now the jaguar is alive, though threatened, in the Maya forest, a tract within the Mesomerican forest of about four thousand square miles of tropical rain forest that extends over the adjoining borders of Mexico, Belize, and Guatemala, where most of Conde's work is focused. The Maya forest comprises a number of national parks and protected zones. In order to save the jaguar, one had to save the forest.

Conde's work was part of a bigger plan to build the Mesoamerican Biological Corridor, which would allow jaguars and other animals to migrate all the way from Panama to southern Mexico. The project was supported by the Central American nations and the investment of $400 million by the Inter-American Development Bank. The problem was the Inter-American Development Bank was also simultaneously investing $4 billion in the construction of more than 332 dams and 4,000 square miles of roads that could, ironically, very well negate the efforts behind the corridor.

Conde was attracted to the jaguar not only for its nobility but because it was a top predator. If you save the jaguar, you also save all the other species that are beneath it on the food web, which are a part of its ecosystem. Plus you save the tropical forest, which is important not only to local species but to North American migratory birds as well.

At least 333 species of birds exist in this region, and the Nature Conservancy estimates that 40 percent of the migratory birds from North America stop in the forests and marshlands of this area during their travels. Natural ecosystems tend to be interrelated.

Jaguars are known to take down a number of medium-size animals including white-tailed deer, smaller local red brocket deer, collared peccary (wild pig), Baird's tapir, agouti (a large rodent), armadillo, and coatimundi (a relative of the raccoon). Jaguars are ambush predators, hunting along paths in the forest, mostly in the night, overcoming their prey with powerful teeth and claws. But in doing this the jaguar is helping the populations of these animals, culling the sick and the weak, a natural process that makes these populations stronger. The predator plays an important evolutionary role in keeping wildlife populations fit and healthy.

Jaguars mostly stay away from people. But they do take an occasional cow, goat, or chicken, possibly putting them at odds with local ranchers and farmers. Conde and other biologists tried to get the governments of Mexico, Belize, and Guatemala to create jaguar insurance whereby they would pay biologists to remove problem jaguars and take them to areas where they would do less harm.

Unfortunately, Conde's studies have been limited due to the costs of jaguar collars ($4,000 to $7,000 each), but the data she has retrieved has given her a vital look into the type of habitat jaguars need. Though the animals would travel through secondary forests and developed land, her collared jaguars spend most of their time in primary or pristine forest. Conde said this showed the jaguar needs undisturbed areas.

Deforestation in Mexico, Belize, and Guatemala is having devastating results. On a cloudy day during the rainy season in Petén, the frontier region of northern Guatemala, I accompanied Conde and Lucrecia Masaya, the research and conservation director at Defensores de la Naturaleza, in Guatemala City, into Laguna del Tigre National Park. According to Masaya, her group was interested in a number of environmental causes and "the healthy populations of jaguars are one way to tell if the things we are doing are working or not," she said.

The dirt road we traveled on was only two years old, yet slash-and-burn agriculture had already destroyed wide swaths of tropical forest along its path. The group took a boat up the Río San Pedro to the Macaw Biological Station. At dusk, we climbed a tower on a nearby knoll, gazed at the surrounding rain forest, watched tropical birds fly by, and listened to the monkeys in nearby trees. The following morning, I accompanied Conde as she showed Masaya a map of the new roads that the government of Guatemala has planned to attract tourism from the Yucatán to the Mayan ruins in Guatemala. The plans called for thirty-nine-foot-wide (twelve-meter) paved roads. Conde referred to the deforestation the group saw on the road leading into the park: "And that was along a dirt road. Can you imagine the devastation that will come from paved roads?"

In the past fifty years, Guatemala has lost two-thirds of its original forested area and the biodiversity that it held. According to the United Nations' figures, since 1990 about 133,000 acres (54,000 hectares) of Guatemala's forests have been lost each year.

The importance of that forest, and how its fate was interconnected with man's, was on display when Hurricane Mitch hit Central America in 1998. The storm formed over the Atlantic and moved toward the central Caribbean Sea in late October. As the storm drifted over warm water, it quickly intensified to a category 5 hurricane with 180-mile-per-hour (290-kilometer-per-hour) winds, then stalled just off the north coast of Honduras, below Guatemala and Belize. The hurricane slowly weakened as it inched southward toward the shore, then westward over Central America. Eventually the heavy rain (36 inches, or 91 centimeters) in Choluteca, Honduras, caused flooding and landslides, killing more than 19,000 people and devastating the entire infrastructure of Honduras and parts of Nicaragua, Belize, El Salvador, and Guatemala. Whole villages and their inhabitants were swept away in torrents of floodwaters and deep mud.

Landslides were particularly virulent on hillsides cleared of vegetation for agriculture. Without the forest to anchor the soils, the rapid runoff from the rains formed rivers of mud. In areas where land had not been cleared, fewer landslides occurred. Even plots of land farmed with crops like coffee and cocoa under the shade of canopy trees did much better than cleared land. Natural and diverse landscapes fared far better than manicured ones.

Mangroves are great buffers against storm damage—more effective than the best concrete dikes, because they capture sediments and build mounds with their roots that keep up with the rise of the sea level. But, since 1950, Guatemala has lost about 65,500 acres (26,500 hectares) of mangrove forests, representing 70 percent of its historic area, according to the Nature Conservancy.

Mangroves can stabilize coastal lands and provide a strong buffer to coastal storms, even hurricanes. Nature has the ability to evolve with change in general, something man does not always appreciate.

For many, Las Vegas, Nevada, with its abundance of neon lights, swimming pools, and wildly decorated hotels might be one place where the concerns of nature could take a backseat to man, but this is not the case. I arrived in Las Vegas after a long day of driving through the desert. I'd come here to see if this neon city ran independently of nature or if its fate was much more intertwined. I checked into my hotel on the main strip and headed out onto Las Vegas Boulevard. It was 11 p.m. on a Thursday, but the city was still very much alive.

The hotels that lined the boulevard looked like amusement park rides. The New York-New York Hotel & Casino was a three-story replica of the New York City skyline and the Statue of Liberty. The Paris Las Vegas had a slightly leaning Eiffel Tower in front of it. The Bellagio looked like Venice, with more than 1,200 dancing fountains that moved to music on a lake of more than 8.5 acres of water.

Charles R. Marshall, an ecologist whom I visited at the University of California, Berkeley, several months before, said, "It's so spectacular, out of control, and extreme. It's one of my favorite places, though that usually horrifies most people I know." Marshall, who grew up in Australia before coming to the US, was married in Las Vegas. His father was, too.

Gambling is king here. My mother once rolled eleven consecutive wins at the craps table at Caesars Palace, and the crowds gathered around the table four or five people deep. They don't get that excited about nature.

Though many visitors see only the man-made side of Las Vegas, it does have a natural history. In the late 1800s, Las Vegas was just a stage stop on the Santa Fe Trail. It had two freshwater springs. Las Vegas is Spanish for "the Meadows." In 1900, the population had grown to around thirty, which didn't even make the census.

But in 1904 the town was picked as the ideal layover spot for crew change and service on the Union Pacific train that ran from Salt Lake City to Los Angeles, and the town started growing. The state of Nevada long embraced permissiveness, and Las Vegas ran with that idea. It allowed gambling, prostitution, quickie marriages, and relatively quick divorces.

Four days before Christmas in 1928, President Calvin Coolidge signed a bill authorizing $175 million for the construction of the Boulder Dam (later rechristened the Hoover Dam) outside of Las Vegas, and the town went wild. Nevada lawmakers made their state the only one in the nation to allow legal, wide-open casino-style gambling. Then they lowered the divorce residency requirement from six months to six weeks and that got Hollywood's attention.

Bugsy Siegel, head of an underworld coalition known as the Syndicate, came to Las Vegas in the 1940s. He was immediately enamored of the whole "Sin City" scenario, built himself the Flamingo hotel, and started palling around with Hollywood stars, including his rumored "old friend" George Raft. But Siegel ran into trouble with the Syndicate, and in June 1947 got a bullet in the eye as a reward.

On January 27, 1951, the Atomic Energy Commission tested the first of a series of atom bombs outside Las Vegas. Soldiers were purposely exposed to the tests to gauge the effects of radiation on human beings. Vegas didn't seem to mind, though the first test left a trail of broken glass across the city. Eventually these tests were moved underground. Over the years Las Vegas has decorated all of its casinos with neon lights—perhaps to make up for the loss of nuclear illumination.

The following morning, I drove a couple of miles off the strip to the University of Nevada, Las Vegas (UNLV), and met with Stan Smith, an ecologist. He showed me some of the desert landscaping that had made the campus famous right outside his office door. The school advertises itself as an arboretum that includes the entire 335-acre campus. Smith had been studying how plants adapt to stress. He'd also looked at how climate change would affect the structure and function of desert landscapes and ecosystems.

An amiable man with wavy silver hair and lots of anecdotes, Smith was raised in Las Cruces, New Mexico, but spent time in Reno, Nevada, and Phoenix, Arizona, before coming to Las Vegas. He was quite familiar with the Southwest desert, although he claimed most Las Vegans were more familiar with the gambling. "You see slot machines all over—at the airport, at the end of the line at the grocers. People in Arizona and California utilize their desert for recreation. But when I was last on jury duty, the other members were comparing coupons from different casinos to see which ones gave the best rewards. Though there are true outdoor enthusiasts here, most people just aren't that interested," says Smith.

Las Vegas casinos keep their curtains closed so you don't look outside. They don't have clocks on the walls, and the lighting is such that it is difficult to tell if it's day or night. Hotels like Caesars have elaborately decorated moving sidewalks to get you inside the casino, but once you are there it is really hard to find the exit sign. And when you manage to escape, it's usually into a parking lot or curbside area that is a lot less friendly than that moving sidewalk you came in on.

* * * *

The outdoors may not impress the majority of its Las Vegas citizens and its fortune-seeking visitors, but nature is the real treasure here. Though the desert shrubs cover only about 20 percent of the desert floor, they are the crucial habitat of lizards, snakes, mice, and birds. Birds and bats are important seed dispersers, eating desert fruits during the wet season and spreading their seeds through droppings. These flowers are essential to the health of migrating birds and raptors. The mountains around Las Vegas contain bobcats, coyotes, mountain lions, desert tortoises, and bighorn sheep. Near Lake Meade on the Colorado River just outside Las Vegas, I stood one hundred yards away from a watering hole at midday and saw twenty bighorn sheep, several with large curling horns, as they came to take a drink.

Though it goes unnoticed by most, among the most important natural elements here are the crusts that cover much of the desert in the Southwest. Biological soil crusts form in open desert areas from a highly specialized community of cyanobacteria, mosses, and lichens. Crusts generally cover all soil spaces not occupied by plants, which can be up to 70 percent of open spaces.

Biological and mineral crusts help keep soil stable, reports Jayne Belnap, a US Geological Survey research biologist in Moab, Utah. A well-developed biological crust is nearly immune to wind erosion. "It's tough as nails against all wind forces," she says. "Tests in wind tunnels of undisturbed crusts in the national parks show that biological crusts can withstand winds up to one hundred miles per hour."

But once these crusts are broken, they become dust sources and can fuel powerful dust storms. That dust can travel quite a distance. Biologists have tracked dust storms over Africa spreading all the way to the Amazon in South America. Dust storms over China have been tracked all the way to the US and out over the Atlantic.

If models of Southwestern responses to climate change are correct, Southwest US deserts should get warmer and drier. With less moisture, crusts may not form, and sandstorms could become much more common. Cyanobacteria, mosses, and lichens are critical to the formation of crusts, and crusts are as important to the residents as gambling, though they don't get much appreciation for their valuable services.

As important as the crusts are, Las Vegas owes its life to the water that is brought to the city by the Colorado River. As with the New York City watershed, the water from the Colorado River originates upriver in less developed forest. The Colorado begins its journey from the snowpack in the central Rocky Mountains and travels south 1,450 miles (2,330 kilometers), draining an expansive yet arid area that encompasses parts of seven US and two Mexican states.

The Colorado River is the principal river of the Southwestern United States and northwestern Mexico. Prior to European settlement, the river entered Mexico, where it formed a large delta before emptying into the Gulf of California off Mexican shores. But for much of the past half century, intensive water consumption upriver has stolen the moisture of the last hundred miles of the river, and it no longer makes it to the Gulf except in years of heavy runoff. Although it is the seventh-longest river in the US, its water volume is quite low. And to make matters worse, for the last couple of decades the population growth along this already strained river has been the greatest in the country.

The immediate outlook is dim, and the long-range picture dimmer. Between 85 and 90 percent of the Colorado River's discharge originates in snowmelt, mostly from the Rocky Mountains of Colorado and Wyoming. Nevada and other Western states like California and Arizona are already struggling with the problem of diminishing snowpack in their own states, and rely on the Colorado River for much-needed water. Climate change will decrease the volume of precipitation in the Southwest while decreasing the snowpack in the Rockies. Water will be released earlier, which means winter and spring may have sufficient moisture but summer and fall will be dry.

The critical part of this equation for Las Vegas and the Colorado River is increasing use by other desert cities, including Phoenix and

Los Angeles. One of the main reasons for building the Hoover Dam in the first place was to bring Colorado River water to Los Angeles and the rest of Southern California, places that never seem to get enough.

Emma Rosi-Marshall, an aquatic ecologist at the Cary Institute of Ecosystem Studies, works with native fish in the Colorado River. The two major dams, the Hoover Dam near Las Vegas and the Glen Canyon Dam below Lake Powell in Utah, have had major effects on wildlife and fish in the Colorado River, altering their natural ecosystems, drowning their habitat, and changing the temperatures of the waters in which they evolved.

Completed in 1963, the Glen Canyon Dam in northern Arizona is one of the last large dams built in America. To provide pressure for power generation, the Glen Canyon Dam draws water from the cold depths of Lake Powell, making the water flowing out of the dam much colder than it is naturally for most of the year. This change in the temperature has had enormous consequences for aquatic species. Worms, snails, and many native aquatic insects have disappeared. These were all-important food sources for native fish. The result is the decline of half the native fishes in the Grand Canyon ecosystem.

Rosi-Marshall works with the charismatic and oddly shaped humpback chub, one of the Colorado River's native fish, a federally endangered species and an important member of the native aquatic environment. Prior to damming, the chub benefited from snowmelt from the Colorado Rockies during spring thaw that would naturally flood the banks of the Colorado River and shape the surrounding wetlands and beaches. With pressure from environmental groups, state water agencies now release water at different times of the year to try to imitate natural runoff. But the benefits of this strategy remain under investigation. It may be that the Glen Canyon and Hoover Dams have altered the river's ecosystem beyond the point where regulating the flow of water through the dams is going to achieve anything like the natural flow of water that existed before them.

The biggest problem for the future of the Colorado River and its surrounding environments is that the river is rapidly losing water, an issue with repercussions for practically every animal and every plant that relies on it, including man. The volume of water in Lake Meade is down to about 40 percent. Las Vegas currently has two major pipes drawing water out of the lake, but the city needs more. Below Lake Meade, the river is drying up. One of the biggest water users is agriculture in Southern California, and UNLV ecologist Smith wonders just how important and productive those farms are. But if you get rid of local agriculture, then you have to go farther away for your food, inevitably putting more CO_2 in the atmosphere from food transport, and that could result in decreased snowpack in the Rocky Mountains and diminished rains in the Southwest desert, causing water levels to fall even lower. As at the craps tables at the nearby casinos, in the end you just can't win.

The Las Vegas Valley, which includes the city, has a population of close to two million, about two-thirds of the people in the entire state. Engineers are proposing to tap underground waters of upstate ranchlands with 145 huge wells spread out over 20 percent of Nevada and connected by one thousand miles of pipe. Such a situation occurred about one hundred years ago when Los Angeles went looking for water in the Owens Valley about three hundred miles upstate on the eastern side of the Sierra Nevada mountains. Los Angeles bought up water rights from Owens Valley ranchers who were misled into thinking they were getting some help with building their reservoirs, but Los Angeles built an aqueduct and sent all the water south.

The Owens Valley slowly surrendered its moisture and the farmers and ranchers moved elsewhere. Water diversions for Los Angeles residents left Owens Lake bone-dry by 1920. Then the dust started blowing. By the 1990s, the Owens Lake playa was the largest producer in North America of PM10 atmospheric dust—particulate matter small enough to enter human lungs. The courts forced Los Angeles to put some water back into the lake, though ecologists continue the fight for more changes in water and land use there. According to

Greg Okin, a professor of geography at the University of California, Los Angeles, "Climate models predict that the Southwest should get warmer and drier, and that by 2050 soil moisture could be lower than the US Dust Bowl Era."

The Dust Bowl occurred in the Great Plains of Midwestern America in the 1930s. An unusually wet period had encouraged people to settle there, and the existing rains convinced many to begin plowing the grasslands deeply. This destroyed the grasses, which normally trapped soil and moisture during times of drought and high winds. Thus when drought came in the 1930s, there was little grass to hold the topsoil. In 1930 an extended and severe drought caused crops to fail, leaving the plowed fields exposed to wind erosion, which carried the fine soils east.

The "black blizzards," as the dust storms were called, began blowing, with disastrous consequences. In May 1934 two dust storms removed massive amounts of topsoil from Great Plains farms and carried it all the way to Chicago, dumping 12 million pounds of dust on that city. Two days later the storm reached the East Coast, dumping huge amounts of dust on Boston, New York City, and Washington, DC, reducing visibility to three feet (one meter) in some places. It has been called the worst drought in US recorded history.

Las Vegas is a human phenomenon, an incredibly large futuristic infrastructure that was built almost entirely in the last hundred years. In 1900 there were about thirty settlers in the valley. Today it has two million. If it took only a hundred years to get to where it is now, how many more years—one hundred? two hundred? three hundred?—will it take to get to the point where there is not enough water for the city to survive, the desert crusts vanish, the dust starts blowing, and the tourists go home?

To get a glimpse of that dusty, thirsty future, all one needs to do is head down the Colorado River to where it ends about fifty miles south of the US border. The water that lies in its bed there is but a shallow, narrow swamp of salt and pesticide-laced runoff from crop irrigation.

Aldo Leopold, an American ecologist, forester, environmentalist, and author of *A Sand County Almanac* (1949), once described the Colorado River Delta as a "milk and honey wilderness where egrets gathered like a snowstorm, jaguars roamed, and wild melons grew." Today, the Cucapá Indians eke out a living in an estuary that is filled with weeds, trash, and occasional swamps of unhealthy water.

Or perhaps the real future of Las Vegas might lie on the banks of the Salton Sea in Southern California, about 120 miles north. This area was born when the Colorado River temporarily diverted into the Salton Sea in 1905. For a time, runoff from farms kept the lake level constant if not polluted. Though the largest lake in California, the Salton Sea is also the lowest, and its water is saltier than the Pacific Ocean.

The Salton Sea enjoyed some success as a resort area in the 1950s as resort communities at Salton City, Salton Sea Beach, and Desert Shores on the western shore and Desert Beach, North Shore, and Bombay Beach on the eastern shore got started and looked promising for a while. But very little development followed due to the area's isolation and lack of local employment opportunities. With no outflow, the lake kept getting more polluted. In the 1970s, most of the buildings constructed along the shoreline were abandoned. The episode "Holiday Hell" from the television series *Life After People* used the Salton Sea as an example of how a resort town like Palm Springs or Las Vegas could decay if there were no humans left to maintain it.

The birds that migrate to the south side of the lake in winter still draw bird watchers, but that is primarily because all the marshlands in the Imperial Valley, where the Salton Sea lies, are taken up by agriculture. There's no place else for the birds to go.

The east side of the sea around the former yacht club is mostly old abandoned trailers and assorted ruins that photographers like to visit—to celebrate what once was, or because some find art in old rusted ruins.

Las Vegas could get there, too. If the water in the soil gets below Dust Bowl levels, the crusts would break down and the sands might pick up and fly with the wind. If the water runs out and the city goes dry, it wouldn't take long for the golf courses, the fountains, and the swimming pools to lose their appeal. And if the desert gets hotter and dryer, the great migration and construction boom of the last fifty years could take its final bow.

Some future artist might revel in the rusted infrastructure of the famous Sin City, go looking for relics of slot machines in the nearby dump, or collect neon artifacts for some museum. Or he or she might go rummaging through old books or magazines to discover the tale of how Sin City finally succumbed to drought, dust storms, and sky-high electric bills, and the day the last neon light flickered out.

Will man's own luck last? Nature holds all the cards.

9

THE LONG RENEWAL

As we've seen, our species is not impervious to the harm we are raining down on the planet. If we keep progressing on all destructive tracks—overpopulation, disease, climate change, destruction of the forests, destruction of the soil, exhaustion of our natural resources—one of them will take us out. Or perhaps it will be the combination of all these factors. We'll go extinct. It's a natural process. Usually it proceeds a little slower. Two hundred thousand years, our current stay on earth, is a short life for a species. When I visited Hans-Dieter Sues, curator of vertebrate paleontology at the Smithsonian, he asserted, "The average mammal species might survive about one million years. A clam species maybe ten million years." But, I tell him, UC Berkeley's Barnosky thinks a mass extinction could come in three hundred years. And Stanford's Jackson says the next hundred years could be crucial. Sues leans back in his chair and smiles resolutely. "Nothing lasts forever," he replies.

Extinction in reality is a simple process. It happens when the death rate of a species exceeds the replacement rate by newborns. This will come for man in five hundred, five thousand, or fifty thousand years as current rates of overpopulation, disease, or all the possibilities listed above continue. Toss in a nuclear war, an asteroid (a regular occurrence in our geological history), or a supervolcano (a major factor in the Permian and Cretaceous), and we're there much faster. One of these could get us, but a multipronged assault will probably yield a cleaner kill.

The problem is we look around at our advanced culture and see an indomitable force. But that's an illusion. We're really more like a virus, about ready to run its course. Said biologist Jim Estes when I visited him at the Long Marine Lab at UC Santa Cruz, "There is no reason to think we will live on in perpetuity when nothing else ever has."

So what if we were out of the picture, not hanging around the old haunts anymore? What would happen to nature? The extinction of *Homo sapiens* would be the equivalent of a soldier yelling for a cease-fire, and the bullets stop whizzing overhead. Nature would be able to catch its breath and calm down, but full recovery from man's 200,000-year assault on nature would take some time.

It took the earth about ten million years to recover from the Permian extinction. It took insects about nine million years to recover from the Cretaceous extinction. Other mass recoveries have been much quicker. How the extinction process might evolve in our current situation is mirrored in earthly catastrophes of the past.

An example of nature's powers for both destruction and renewal were on display on the morning of May 18, 1980, when the entire north side of Mount St. Helens in the state of Washington collapsed as forces from the interior of the volcano exploded through its cauldron. The blast took the lives of fifty-seven people, including Harry Randall Truman, owner and caretaker of Mount St. Helens Lodge on Spirit Lake at the foot of the mountain. Truman had stubbornly refused to leave his home despite numerous warnings.

The explosion toppled most of the trees in an area called the "blowdown zone" that stretched north over 143 square miles; the trees now on the ground all pointed away from the blast like fallen soldiers. Trees at the edge of that zone were scorched and killed by the flow of superheated rock and gas that shot from the volcano's mouth at 125 miles (200 kilometers) per hour, searing the landscape with material up to 1,200 degrees Fahrenheit (650 degrees Celsius). It created a barren plane of pumice up to 131 feet (40 meters) thick stretching out six miles to the north.

Most of the lands in the blast area were on private or forest service property and were therefore part of an extensive salvage logging operation that had the forest replanted and growing within five years. But that wasn't the case on the 110,000-acre Mount St. Helens Volcanic Monument established by the US Congress in 1982 to follow the natural return of the forest from the eruption. Researchers claim that Mount St. Helens is today the most studied volcano in the world.

I visited the park ten years after the eruption and spent most of my time chasing several herds of elk, trying to get pictures of the species' return to the area. Grasses and smaller plants had moved into the vacancies created by the fires that followed the blast. Elk from outside the park moved in and took advantage of these shoots even if they were buried under ash.

More recently, the monument celebrated its thirtieth anniversary, and monument biologists and geologists are recording nature's process of renewal. Wildlife had a couple of breaks at 8:32 a.m. on May 18, 1980, that sped the recovery. Spring had come to the mountain late that year, so there were snowdrifts on the ground that protected the brush and plants of the forest and the animals beneath them. Lakes were still frozen and many fish and amphibians survived intact under the ice. Since this all occurred in spring, migratory birds as well as the salmon had yet to return. Nocturnal animals had already bedded down by the time of the explosion, some in burrows, and they fared far better than their wide-awake neighbors who were up at the crack of dawn.

Plants started to return in the first few years, their seeds emerging from the ash or carried in by reinvading animals. Wind has also played a key role in blowing in spiders, insects, and seeds. Prairie lupine, a mostly purple or blue wildflower with soft silvery green leaves, came back even on pumice within a couple of years. The plant, which fixes its own nitrogen from the air, created small microhabitats for other plants. These plants trapped windblown debris and attracted insects, all of which ended up enriching the soil beneath them with organic matter.

New grasses and plants provided food for the birds, small animals, and larger plant eaters that followed. Ten years after the blast, the most common large animal species were elk, black-tailed deer, mountain goat, black bear, and cougar. The elk and the deer were most prevalent. Resident populations at the time of the blast that were too big to get out of the way or hide in burrows were cut down mercilessly. But open habitat created by the volcano and the fresh plants that emerged have attracted other populations of animals from outside the park back into its boundaries.

Plant colonization had occurred in boom-and-bust cycles. Single species rushed into areas that were free of competition—new grasses being the most dominant here at first. These grew explosively, but once predators, parasites, and competition returned, the newcomers tended to crash. Gradually though, more species established themselves and diversity returned. And with that diversity came stability as stable communities established themselves and the pace of succession slowed down.

When I visited the park thirty years later, the forest had revived in patches where trees were buried under snow or protected by rocks. The species makeup of the surviving forest had changed in spots. More shade-tolerant, understory trees like mountain hemlock had emerged to dominate the landscape where the taller Douglas fir trees would have grown before the eruption. The layer of ash that fell from the sky that day killed trees such as Pacific silver firs even years after the explosion.

However, conifers, the dominant trees in the Cascade Range where the volcano is located, had not returned in force. They were susceptible to drought and needed a certain type of fungus in the soil to help them grow. The succession of forest growth after fire or fiery volcano was a series of vegetation changes with brush or less stable trees coming in first, followed by more dominant, stable (what scientists call "climax") vegetation coming in last. Conifers, trees that bear cones—like Douglas fir, western hemlock, lodgepole pine, and Pacific silver fir—will dominate perhaps in the next few decades, but it will be hundreds of years before a true old-growth forest will reappear.

The eruption of Krakatoa, an Indonesian island between Sumatra and Java on August 27, 1883, is another example of nature's propensity for destruction and renewal. It is often referred to as the first great natural catastrophe of the modern world, since telegraph wires had recently been laid across the oceans, and the explosion became international news at the speed of electronic transmission.

The volcano had been sending up churning clouds of ash and pumice along with explosive noises for almost two months. Villagers in the surrounding islands greeted these natural fanfares with near-festive activities. But no one was prepared for what came next: one of the largest eruptions in modern times.

The series of cataclysmic explosions began at midday on August 26 and lasted until the next day, ending with the grandest explosion of them all. On that second day, the northern two-thirds of the island collapsed beneath the sea, generating a series of huge eruptions, followed by a series of tsunamis that raced toward the surrounding islands. The waves lifted boats into the air and swept whole villages out to sea. The death toll was more than thirty-six thousand people.

A police chief on Rodriguez Island could hear the enormous bang of the volcano, "like naval gunfire," though he was 2,970 miles (4,770 kilometers) away—the equivalent of someone in London, England, hearing an explosion in Baltimore, Maryland. The tower of ash and pumice rose to a height of nearly 30 miles (48 kilometers), raining down huge masses of pumice on the surrounding seas. Some islands of pumice were found later, floating in the water, laden with skeletons.

So what has happened to the area since the eruption? The story is encouraging: within a century the remnants of Krakatoa, where not a blade of grass was visible for a year, were draped in tropical forest from sea level to the 2,600-feet (800-meter) peak. There were now over four hundred species of plants, thousands of species of arthropods (spiders, crustaceans, and insects—including fifty-four species of butterflies), more than thirty species of birds, eighteen species of land mollusks, seventeen species of bats, and nine species of reptiles, many of which had to cross forty-four kilometers of sea water to even reach the islands. No species count exists prior to the eruption, but the numbers of animals challenges other counts in nearby areas.

Professor emeritus Ian Thornton of La Trobe University in Australia, who wrote many papers on the volcano, reported that Krakatoa offered "an optimistic lesson: That tropical rainforest ecosystems are capable of recovery from extreme, traumatic damage, if left alone and given time."

To scale up from the aftermath of the eruption of Mount St. Helens or even the eruption of Krakatoa to the aftermath of the Permian extinction is an enormous leap. Still, some of the same principles apply.

The landscapes that existed at the Permian extinction were also barren. The Siberian Traps had spewed enough volcanic matter to cover an area the size of the continental United States. The coming together of all the continents into the supercontinent Pangaea had shut off ocean circulation, and life in the deep oceans began to lose oxygen. Stagnation started to replace moving currents and the result was the release of sulfur dioxide (SO_2), a deadly poison. The oceans had gone acidic, and shellfish and coral couldn't grow hard shells. Unlike Mount St. Helens and Krakatoa, the buildup to the Permian extinction was not a singular eruption but a series of eruptions with consequences that lasted thousands of years.

The earth didn't rebound rapidly. Early Triassic rocks are notoriously barren of fossils, making it hard, researchers claim, to get grad students to work them. Douglas Erwin, in his book *Extinction*, compares the earliest Triassic to the ravages of the Scythian hordes in *Prometheus Bound*: "This is the world's limit that we have come to; this is the Scythian country, an untrodden desolation," says Aeschylus.

The recovery, as at Mount St. Helens and Krakatoa, came from survivors and immigrants that surrounded the catastrophe: the birds, fish, small mammals, and reptiles. The larger animals—the elk, deer, coyotes, and mountain lions that followed Mount St. Helens—came later. In the presence of a vacuum, evolution advanced the cause of life as it did with the spread and multiplication of new species of finches that first populated the Galápagos.

With the Permian extinction, the destruction was greater, and the recovery lasted longer. Douglas Erwin, paleobiologist at the Smithsonian, compares the Permian recovery to an empty chessboard where each square represented unique ecological niches. The empty spaces that followed the extinction presented different opportunities for life, for rapid speciation and expansion of individual territories. Those circumstances were more complicated than at Mount St. Helens or Krakatoa—whose eruptions did not lead to new species—since the Permian extinction led to wholesale species changes, and it took ten million years (not one hundred years) for things to start coming back together.

For the Permian extinction, it wasn't just a matter of animal species lost. Many of the rules that governed ecological relationships were abandoned. According to Erwin, the board collapsed entirely, and as the game resumed, it became "half chess, half backgammon, with some rules drawn from poker." During the Permian, only one in ten snails or slugs was a predator. They made their living searching through the water and mud for organic debris. Some of them were grazers chewing on algae. But after the Permian extinction all that changed, as slugs and snails became vicious predators, many equipped with highly toxic poisons to capture prey.

NEW SPECIES

The landscape of the first three million years of the Triassic was like a ghost town. The tiny fraction of species that were alive then was but a small portion of the species that thrived a few million years earlier, or even a few million years later. Before the extinction, passive groups of animals dominated, but after the extinction, active groups took control. Sitting around, hoping your food would come to you, did not work as well as going out and getting it in this changed environment.

According to Smithsonian paleontologist Hans-Dieter Sues, the resurrection of the environment came in fits and starts. On land, the number of species lost was not as great as the loss in the oceans. Some species reappeared from refuges, safe havens for relic plants and animals. The tropics may have been such a safe haven.

An extreme example of these refuges can be found at Fray Jorge National Park in northern Chile. Upon driving into the park, one can see only desert. This area receives less than six inches of rain a year, and the desert shrub is more suggestive of the badlands of the American Southwest than the lush landscapes of the Amazon. Yet perched atop the coastal mountains, some 1,500 to 2,000 feet (460 to 600 meters) above the level of the nearby Pacific Ocean, are patches of vibrant rain forest extending up to 30 acres (12 hectares) apiece. Trees stretch as much as 100 feet (30 meters) into the sky, with ferns, mosses, and bromeliads adorning their canopies. After leaving your car, you climb up an arid desert path and discover that it turns from dry desert shrub into forest. And then the biggest surprise: as you enter the forest it suddenly starts to rain.

This is not rain from clouds in the sky above but from fog dripping down from the canopies of trees—trees so efficient at snatching water out of the air that they get three-quarters of all the water they'll ever need from the fog. That same fog at Fray Jorge also provides nutrients. Kathleen C. Weathers, a biogeochemist at the Cary Institute for Ecosystem Services, and her colleagues have discovered that this fog, originating offshore from some of the richest ocean waters on the planet, floats in bearing essential nitrogen and other chemical gifts. Similar bizarre sanctuaries may have safeguarded species during the Permian extinction.

During the early Triassic, most of the interior of Pangaea was a hot, dry desert. The continental plates that made Pangaea were fused together, but as Pangaea finished its final assembly, the plates began to rip apart again. By the end of the Triassic, North America was pulling away from Europe and Africa as the crust between them sank, forming the Atlantic Ocean. Still, areas of the southern continents remained in tropical forest—safe havens, perhaps, like the fog forests of Chile.

But then species began to develop at greater rates than ever before. The first species attracted to the barren landscape of the early Triassic, as with Mount St. Helens and Krakatoa, were the weedy opportunists, "the ecological equivalent of dandelions springing up unbidden in a spring lawn," writes Doug Erwin in *Extinction*. Though these "weeds" weren't always plants, they acted like weeds in the sense that they moved into open territories and proliferated. The piñon pine forests of southern Utah bear the calcified remnants of the early Triassic scallop *Claraia*, a weedy species whose fossil shells today form pavements built from the remains of thousands and thousands of mollusks who thrived here when much of the western United States was covered with ocean.

Ferns were some of the first major colonists in other areas. They formed areas similar to the savannas or grasslands of today. Conifers were probably the first large trees of the Triassic. Most of the petrified trunks seen at Petrified Forest National Park in Arizona are conifers. They shared the landscape with tree ferns as well as ginkgos, which are related to conifers. The only member of the ginkgo genus that survives today is *Ginkgo biloba*, which has been used in Chinese herbal medicine for many centuries. The tree has fan-shaped leaves. The Japanese sometimes call it *I-cho*, "tree with leaves like a duck's foot."

Barren land free of all vegetation gradually began to disappear. As the search for space grew more competitive, plants began to move into the lowlands, forming swamps, which led to coals reappearing. But it was not until the late Triassic that the earth was covered with green again.

* * * *

In the wake of the Permian extinction, plants and animals went through similar successions. Hans-Dieter Sues led me into a back room in the Smithsonian Institution one spring day to show me some important animal fossils that thrived in the world of changes that occurred in the aftermath of the Permian extinction. The room was filled with boxes stacked on metal shelves. Sues, smiling, reached into one box and pulled out a skull. "This is *Lystrosaurus*," he said, as if it were a friend. "They were the most dominant group of animals of the early Triassic era, even if they were rather ugly." The skull is round, with a pug nose and sockets where the tusks had once been. It wouldn't sell beer or toothpaste.

The return of animals during that time mimics the return of plants. In the early Triassic there were few species of animals, but those that survived spread throughout the world. Fossils of *Lystrosaurus* are the most numerous remains from the Permian extinction. It was a relative of early mammals. Its skin was smooth like a hippo. A horny beak may have covered the upper and lower jaws. Some cousins of *Lystrosaurus* reached body weights in excess of 2,200 pounds (1,000 kilograms). Still, he got around. He was the dominant vertebrate animal in the early Triassic in South America, India, Antarctica, China, and Russia.

There were few predators in the early Triassic. Most animals were dead and there wasn't enough for a good predator to eat. The only real consumption going on was by fungi that were attracted to all the dead bodies. Sues pulled another fossil out of the box. This was the skull of a gorgonopsid, a nasty predator with huge canine teeth. He was the dominant predator of the late Permian. Some of his relatives attained the size of a lion. But the gorgonopsids didn't make it across the Permian-Triassic extinction line.

The sheer intensity of the extinction, plus a decrease in oxygen, warming, and other crises that continued in lethal bursts for five to six million years after the Permian extinction, delayed the resurrection in the early Triassic. Even ten million years after the Permian extinction things were bleak. River drainage patterns confirm a catastrophic loss of vegetation, which didn't bode well for plant eaters. Global warming, acid rain, ocean acidification, and ocean anoxia (the absence of oxygen) continued for a while, as did the greenhouse gases that precipitated them. Greenhouse gases don't vanish quickly. There was life then, but it was meager. Then, 240 to 230 million years ago, things started to change.

This is when the crocodylomorphs (crocodile-like animals) and the first dinosaurs started to form. In the seas, ancestral crabs and lobsters as well as the first marine reptiles were creating some of the first ecosystems, but the crocodylomorphs weren't semiaquatic beasts like their modern-day progeny—crocodiles, alligators, caimans, gharials—they were terrestrial beasts. And they weren't the second-best predator of the day. They ruled the lands, the most vicious predators on earth.

There is a rendering in Hans-Dieter Sues and Nicholas C. Fraser's book *Triassic Life on Land: The Great Transition* that gives you an inkling of the ferocity of the crocodile-like animals during the late Triassic. It is a scene from western North America. A colossal phytosaur, looking like a diesel truck only with a long thick tail and a mammoth gaping crocodile mouth, is standing in a shallow area of water, surrounded by animals with tall spindly legs, looking like a pack of dogs but with that same crocodile-like face. There's five of them, and although the figure of the phytosaur is scary even as a drawing, the crocodylomorph reptiles have completely surrounded him and are not as impressed with his size.

For the most part these were dry-land creatures. The shape and musculature of the jaw distinguished them from other animals. During the Triassic, crocodylomorphs spread across the lands, evolving into different forms, from slender, long-legged, wolflike animals to huge, fearsome animals that were the apex predators of the food web (formerly food chain).

At the end of the Triassic, about 200 million years ago, more volcanic activity in the Central Atlantic elevated CO_2 in the atmosphere with some of the same results that had occurred in the Permian, and the crocodylomorphs lost their advantage to the dinosaurs as many of the largest croc species died out. With the land cleared of competitors, the dinosaurs expanded their dominion, evolved into different species, and took over.

But the crocodylomorphs didn't lose all their ferocity. Paul Sereno, a paleontologist at the University of Chicago, discovered a number of prehistoric crocodylomorphs living alongside the dinosaurs in the wetlands of the ancient Sahara 100 million years later. They were still fearsome creatures.

Sarcosuchus imperator, nicknamed SuperCroc, was some forty feet long and weighed eight tons. What Sereno refers to as BoarCroc was twenty feet long and had three rows of fangs, what Sereno refers to as a "dinosaur slicer." But it also had long, agile legs rather than the squat, close-to-the-ground legs our present-day crocodiles possess. While modern crocodiles wait by the water and leap out to grab their prey, BoarCroc could have leapt out of the water and charged up the bank after dinosaurs.

Crocodile-like animals were the dominant predator to evolve during the 50 million years of the Triassic that followed the Permian extinction. The defenses of the plant eaters grew stronger, as did the attack mechanisms of the new creatures that preyed on them. Crocodiles ruled for a while, but then came the dinosaurs, perhaps the most successful creatures to evolve over the last 600 million years. Even during their reign, mammals were hiding out in the bushes, waiting their turn to take over.

The litany of these creatures proves two things: that even the strongest of animals are vulnerable; and that, though the characters, or the species, may change, life goes on.

A mass extinction may have grave consequences for some species, ours included, but it will not stop life. In the form of plants, animals, birds, reptiles, fish, fungi, and bacteria, life will find a way to exist, and will eventually adapt to survive any conditions thrown at it by man, natural selection, or the universe. Evolution has proven for over the last three billion years to be unstoppable. Nature survives even in war zones. If you give nature space, it finds a way to persist.

On a crisp winter afternoon, Dave Choate, a researcher from the Orange County Cooperative Mountain Lion Study, and I stood at the top of a high hill in the Santa Ana Mountains in California listening for beeps on his tracking mechanism that told Choate there were mountain lions from his study nearby. We were surrounded by the Camp Pendleton Marine Corps base, where 175,000 men and women train each year. Overhead, a squadron of fighter jets streaked toward the bombing range. At various times we heard the sounds of mortars, machine guns, and exploding rockets.

Yet, in spite of the noise, about 75 percent of Camp Pendleton is a de facto wildlife refuge—a huge military reservation comparable in size to the state of Rhode Island. The military needs open space to train service personnel, just as artillery and planes need buffer zones around the ranges where they direct their shells and bombs. If we were to fly overhead, we would see an area pockmarked by military activities, but we would also see a checkerboard pattern of civilian housing and shopping malls that virtually surround the base.

Mountain lions, bobcats, coyotes, and badgers hunt deer, rabbits, and rodents in areas where hawks, falcons, and eagles fill the air along with a multitude of ducks and shore birds. The base even has a herd of buffalo. Military lands are well patrolled, so there's little poaching. Military laws deal out harsher punishments for violators. Says one US Air Force colonel, "We're kind of mean SOBs if you break the rules."

Driving south past Camp Pendleton on Interstate 5, the coastal highway, you leave urban sprawl and enter an area that, for all its failings, is open mountains shrouded in golden grasses and coastal shrubs confronting long sandy beaches. Big oak trees with branches that fall to the ground, the result of a paucity of grazing animals, punctuate the many fields of herbs and flowers. If it were not for the marines, this could all be houses, gas stations, and mini-malls.

* * * *

The Korean demilitarized zone is another example of how nature can hang on under the worst circumstances. The zone is a 148-mile (238-kilometer) line that bisects the Korean Peninsula at the 38th parallel. It represents the armistice boundary between North and South Korea that was established in 1953, after several years of war between the two states. A ten-foot (three-meter) chain-link fence topped with razor wire prevents combatants from going at each other.

The armistice stopped the carnage (almost 900,000 soldiers and 2 million civilians killed or wounded as of July 27, 1953) but not the conflict, as the two states are technically still at war. Hundreds of thousands of troops from two large armies and more than 30,000 US troops stationed in South Korea patrol the area armed with live bullets, backed up by tanks, artillery, and ballistic missiles, all on alert.

But this brandishing of weapons can't take away from the value that this no-man's-land provides for nature. Wetlands created by five rivers, and the steep, forested Taebaek Mountains, make this place the perfect wildlife sanctuary.

The DMZ's roughly four hundred square miles are home to musk deer, black bears, and lynx. About one-third of the world's population of red-crowned cranes depends on the DMZ for habitat. Ninety percent of the planet's black-faced spoonbill population breeds on islands located here. And approximately 1,500 of the earth's largest vulture species, the black vulture, winter here as well.

The loss of the DMZ would bring ruin to populations of goatlike Amur gorals, Siberian musk deer, and other Korean animals, according to Ke Chung Kim, of Pennsylvania State University and cofounder of DMZ Forum, which advocates for the protection of the DMZ as a peace park. Right now nature is being protected by one of the largest and most well-armed military guards in the world. Other such zones have been created during past conflicts, including the United Nations buffer zone between Iraq and Kuwait, and the Vietnamese demilitarized zone between North and South Vietnam, all fine examples of what nature can do in the absence of man.

Perhaps the primary example of nature's long-term survivability is the exclusion zone around the former Chernobyl nuclear power plant in Ukraine. It has been more than a quarter century since the Number Four reactor exploded at the plant. Then dangerous radioactive material spread over vast areas of Ukraine, Belarus, and Russia. Today whole towns are still abandoned. Cancer rates from people in the surrounding areas are high. But the 1,100-square-mile (2,850-square-kilometer) exclusion zone created around the failed reactors is home to a surprising number and variety of wildlife.

Roe deer and wild boar wander among the deserted villages, while bats fly in and out of vacant houses. Wild boars have also taken a liking to the villages. Rare species such as lynx, Przewalski's horses, and eagle owls are thriving in areas that people have abandoned. Even wolves have made a comeback here.

It's not that all is peachy, no problems, nothing to worry about. James Morris, a University of South Carolina biologist, works in the "Red Forest" (so named because the pine needles all turned red after the reactors went down). He's seen trees with weird, twisted forms, the result of radiation destroying the trees' ability to know which way is up and which is down.

A study in the *Journal of Animal Ecology* shows that reproductive rates are much lower in the Chernobyl birds than in control populations. Another study in PLOS (Public Library of Science) says that the brains of the local birds are 5 percent smaller than average and that this may inhibit their survival. Around 40 percent of the barn swallows return each year in other areas, but the annual return rate at Chernobyl is 15 percent or less.

Yet a recent study by Professor Jim Smith at the University of Portsmouth, UK, says that most wildlife has recovered from the initial radiation problems, and that they are doing better than before simply because the human population has been removed. Kiev ecologists believe radiation effects will diminish over time, but that the real story is how Chernobyl has burst into life. They hope one day the area can be turned into a national park.

Chernobyl is not our only nuclear problem. On March 11, 2011, the Tōhoku earthquake and resulting tsunami swamped the Fukushima Daiichi nuclear power plant, located 149 miles (240 kilometers) north of Tokyo, in Fukushima prefecture (province), and cut the power to vital cooling systems for three reactors in use. The result was the second-worst nuclear accident in history. The facility remains toxic to this day.

Censuses of wildlife at Fukushima found that the abundance of birds, butterflies, and cicadas had decreased; bumblebees, grasshoppers, and dragonflies were not affected; and spiders actually increased in abundance—possibly because the insect prey they normally fed on were weaker and easier to catch. Eventually insects will start to drop off. Small mammals, reptiles, and amphibians remain quite low, but cleaned-up areas of the exclusion zone could start attracting them later. Scientists believe that mutations will appear as insects and animals cycle through more generations.

What concerns some biologists is the radioactivity that washed into the ocean. Japan is on the migratory route of multiple marine species in the North Pacific, including tuna and sea turtles. Right now the Fukushima accident site has the stronger effects of the initial explosion and release of radioactive, short-lived isotopes, whereas some of those initial effects of radiation at Chernobyl have disappeared.

Guns, bombs, and radioactive waste aren't the greatest things for wildlife, but they are better than burgeoning populations of humans consuming every last inch of open space. They keep people, concrete, and asphalt at bay by giving the plants some ground to grow on. Still, man has many less obvious ways of destroying wildlife habitat: Just look to our oceans.

10

TROUBLED SEAS: THE FUTURE OF THE OCEANS

THE OCEAN COVERS 71 percent of the earth's surface and contains 97 percent of the planet's water. There are massive amounts of energy stored up in its ponderous waves, occasionally unleashed by storms and earthquakes. Life got started beneath its surface, and it still offers an elegantly evolved storehouse of creatures within its churning waters, though its abundance was greater before we built boats and headed out to sea.

Despite its enormous significance to humans, we know as much about this underwater world as we do about Mars. The ocean is a no-man's-land of weakly controlled international agreements. It's the last frontier. The last place where we still hunt wild game in significant amounts. The last place we still harvest wild creatures with only rudimentary ideas about their limits.

Overharvesting the sea is not our only dilemma, as we must also deal with a legacy of pollution festering within it. The oceans of the world are beginning to absorb the increasing levels of CO_2 we harbor in our atmosphere. This heightens the acidity, and lowers the pH of ocean waters, which is bad for krill, the preferred food source of a number of whales that feed in the high-latitude areas of both the Arctic and the Antarctic. But it gets more exact. Biologists also believe that ocean acidity decreases the ability of whales to hear the mating calls of others. Both of these effects could be catastrophic for whale populations.

I got to see firsthand the importance of human changes to the marine environment when I accompanied Adam Pack, associate professor at the University of Hawai'i at Hilo, aboard the *Kohola II* to the mating and calving grounds of the humpback whales off the Hawaiian Islands. Just a few minutes out of Lahaina Harbor on the island of Maui, a huge humpback whale leapt out of the water, its entire body hanging in the air momentarily before crashing into the ocean, the spray soaking all the researchers on our boat.

But Pack's attention wasn't focused on the breaching behemoth. Instead, he was observing a group of whales roiling at the ocean surface farther off. Soon our boat was just outside a ring of male humpback whales surrounding a lone female.

Pack, dressed in a wet suit, slipped over the side of the boat with a video camera while several of his students and I watched from above. More than ten thousand humpbacks migrate annually from their winter feeding grounds off Alaska and the North Pacific Rim to Hawaiian waters. Surrounding this lone female, the males butted heads and slashed each other with their fins in an effort to position themselves next to her as the principal escort—the one who gets to mate with the female whale first when she becomes receptive.

Studies by Pack and colleagues have shown that the larger females prefer larger males. On this day it seemed that fewer than half of the males swimming in this dangerous circle were juveniles. Still, this is a lot when you consider how much they sacrifice to get here, and how little they are rewarded. Juveniles come here to watch; they don't have access to the females. They are basically traveling six thousand miles from southeastern Alaska to Hawaii—a trip that will cost most of them one-third of their entire body weight due to fasting—to attend a very expensive school on mating behavior.

Biologists aren't quite sure why these whales make this long trek. It may be that Hawaiian waters are warmer and calves don't need such a thick layer of fat around them at birth. Or perhaps it is the fact that there are fewer predators, particularly killer whales, in Hawaiian waters. A study by John Calambokidis with the Cascadia Research Collective, in Olympia, Washington, found that more than 25 percent of humpbacks examined had tooth marks on them from killer whale attacks. But humpback whales take these risks for the chance to mate.

Males advertise themselves to females not only with their size but also with their song, an important part of reproduction. Though juveniles are excluded from mating, they still get to sing. Scientists at the Australian Marine Mammal Research Centre in the 1990s recorded two males singing a particular song one year that was different from the other eighty singers recorded off Australia. And the next year more males were singing that song. The following year all the males were singing that song. A couple of whales had started a musical trend, a form of culture. But ocean acidification may be affecting their song as well as their food.

Humpbacks are adaptive animals. Researchers at the Alaska Whale Foundation have witnessed humpbacks diving below schools of krill or fish and blowing bubbles around the schools, essentially herding them into a tighter group, after which the whale comes up beneath the group, its mouth open wide to capture everything possible.

At one time scientists held the idea that the ocean might be a legitimate sink for growing amounts of CO_2 on land. Some scientists were even looking for ways to improve the uptake of CO_2 by the sea, but it turned out the ocean was doing a good job of taking in CO_2 all by itself. CO_2 in the ocean reacts with the water to form carbonic acid, and this leads to increased ocean acidity. The result is that the oceans are 30 percent more acidic than before. And there are consequences to pay.

Acidification of ocean water is bad for krill, the preferred food source of a number of whales. Studies from the Australian Antarctic Division show that most krill embryos exposed to high levels of acidification (2,000 parts per million) did not develop and none hatched successfully. Cold waters absorb more CO_2 than warmer waters. Southern ocean carbon dioxide levels could rise to 1,400 parts per million by the year 2100, three and a half times higher than current rates closer to the equator. This could devastate marine life.

Ocean creatures that wear their skeletons on the outside, such as shrimp, clams, and coral, will find that an increasingly acidic environment could start dissolving those shells. Krill look like tiny shrimp whose skeletons are wrapped around their bodies like a thin suit of armor. These exoskeletons protect them from the elements, but ocean acidification could destroy that protection.

Plus, acidification interferes with the ability of whales to hear others sing. Researchers at the Monterey Bay Aquarium in California found that acidification reduces the ability of the sea to absorb low-frequency sound. This amplifies the ambient noise level from currents, animals, and man, making it more difficult to hear whale sounds, which are broadcast at similar frequencies. The ocean absorbs at least 12 percent less sound now than it did in preindustrial times. And this is projected to rise to 70 percent in 2050. As the ocean gets noisier, whale sounds may get muffled—a critical component of their mating system.

Humpbacks and other whales evolved from the same terrestrial animals that gave rise to sheep and deer. About 60 million years ago these animals moved back into the sea, slowly evolving the ability to drink salt water as their nostrils moved higher up their foreheads until they became blowholes. Their ancestors spawned different lineages of marine mammals, including whales. Some, like killer whales, preyed on different marine mammals, including other whales; others, like the humpback, evolved fine, fibrous combs called baleen in their mouths to filter shrimp, krill, and other creatures that traveled in large schools.

Though originally from the ocean, they were unable to get their gills back: "Evolution doesn't move backwards," said Hans-Dieter Sues when I visited him. So whales had to learn to breathe air only at the surface. They gradually lost their legs, though some whales still have small vestiges of legs near their tails. That any animal could go through such an enormous range of changes is testament to evolution's incredible ability to morph its creatures.

Commercial whale hunting in the first seven decades of the twentieth century reduced their numbers by over 99 percent. From prewhaling estimates of 250,000 animals, humpback whales had been nearly hunted to extinction, with only about 2,000 then remaining. In 1970, they were put on the endangered list, and since then humpback numbers have rebounded to more than 20,000 in the North Pacific. But acidification could change that progress, particularly since acidification goes hand in hand with warming (both caused predominantly by CO_2).

Warming could result in a loss of polar ice. Some biologists refer to it as the "Atlantification" of the Arctic. A loss of sea ice could affect Arctic whale natives like ivory-white belugas and the single-tusked narwhals, which look like unicorns. These two whales lack a prominent dorsal fin—the main fin located on the back of fishes and certain marine mammals—which makes it easier for them to hunt under the ice. But as the ice cover melts, killer whales—whose prominent dorsal fins have foiled their ice cap hunting so far—could have free rein over the Arctic natives. Killer whales might target bowhead whale calves, while minke whales could provide increasing competition for food to all.

The prospect of a polar-ice-free future concerns many researchers. Gretchen Hofmann, a marine biologist at the University of California at Santa Barbara, makes annual visits to McMurdo Station in the Antarctic to study the effects of acidification. She likes to go down in the southern hemisphere's spring, and she told me: "There are twenty-four hours of daylight but the ice is still strong enough for us to move around on it and support our weight."

McMurdo is a coastal station at the southern tip of Ross Island, about 850 miles (1,360 kilometers) north of the South Pole. It is a snow-covered island surrounded by frozen seas and rimmed by jagged mountains. The annual temperature is zero degrees Fahrenheit (minus 18 degrees Celsius), but it can get even colder with the windchill factor. Many scientists wear "ice cream suits"—big, thick coveralls that cover the whole body—but Hofmann likes her layers better: a down jacket, topped by a layer of polar fleece, topped by another layer of polar fleece, topped by a parka to cut the wind.

She says the worst thing about McMurdo is the food. "It's all from cans. You get used to having fresh vegetables in Santa Barbara. But down there, there's nothing fresh, and your food habits get worse and worse. All of a sudden you realize, 'I'm living on Pringles!'"

Hofmann spends about a month or two each year doing her research and teaching classes. She claims the Antarctic is a special land of snow and ice, but the poles are more affected than other areas by acidification as well as global warming, because colder water holds more CO_2. Hofmann also works in the South Pacific island of Moorea and along the California coast.

Off Antarctica and off Palmyra Atoll in the mid-Pacific, Hofmann and her coworkers have found that the increase in seawater acidity caused by greenhouse gas emissions is still within the bounds of natural pH fluctuation. But areas in California such as at the mouth of the Elkhorn Slough in Monterey Bay and off La Jolla, at the top of San Diego Bay, are already experiencing acidity levels that scientists had expected wouldn't be reached until the end of the century. Hofmann believes ocean acidification in the open ocean may still be tolerable for marine organisms, but that those animals living in tidal, estuarine, and upwelling regions may be functioning at the limits of their physiological tolerance.

Curt Stager, author of *Deep Future: The Next 100,000 Years of Life on Earth* and a professor at Paul Smith's College, has studied the Eocene climatic optimum, an interglacial period that began about 50 million years ago. During this time average global temperatures rose 18 to 22 degrees Fahrenheit (10 to 12 degrees Celsius) above today's mean temperature for several million years.

But what interests Stager most is a brief spike in rising temperatures, called the Paleo-Eocene thermal maximum (PETM), that for approximately 170,000 years forced this world into an extremely warm state, another 10 degrees Fahrenheit (5 to 6 degrees Celsius) hotter, on top of an already warming world that resembles our own extreme-emissions scenario in climate models. To date, humans have sent 300 gigatons of fossil carbon into the atmosphere. During the PETM there were at least 2,000 gigatons in the atmosphere from causes that yet remain unclear.

As greenhouse gas concentrations rose, they warmed and acidified the deep sea enough to wipe out bottom-dwelling creatures and burn a red layer into the ocean floor. Sediment cores show that it took thousands of years for the worst of it to subside. The PETM might have reduced the nutritional value of plants, stunted the growth of mammals, and encouraged insects to attack plants more vigorously. During the PETM, mammals were extremely small, about half the size of their counterparts during the periods before and after.

Increased CO_2 in the bloodstream can reduce an organism's ability to bind and transport oxygen, which is perhaps one of the reasons for the appearance of PETM dwarfs.

Such a high CO_2 scenario would have enormous effects on our present-day coral reefs. Coral reefs are breeding grounds for fish, but with acidification, corals don't aggregate or form stony structures for other marine creatures to cling to or crevices in which to hide. Coral reefs are natural breakwaters for many South Sea islands. But acidification and sea level rise are threatening these places.

Maria Cristina Gambi, of the Stazione Zoologica Anton Dohrn, in Naples, Italy, studies natural volcanic CO_2 vents off the island of Ischia in the Gulf of Naples. She and her colleagues have found fewer animal groups and lower biomass in the extreme low-pH areas near the vents. Instead, a few small acidification-resilient species have filled the gap with population booms, which decreases the number of species.

During the Permian, ocean acidification left a unique legacy in its sedimentary layers, the "Lazarus taxa"—"taxa" meaning biological groups. Certain species seem to disappear at the end of the Permian but then resurface millions of years later, apparently coming back from the dead, as Lazarus did in the Bible.

The resurrection of these creatures may be due to ocean acidification. Without a shell or an exoskeleton, many creatures would leave no fossil or other evidence of their existence. It could be that many of these creatures survived "in the nude" for a while and came back when the oceans were less acidic and more hospitable to building shells.

Mary L. Droser, a paleontologist at the University of California, Riverside, believes Lazarus taxa may actually represent not a resurrection of old species but the convergent evolution of other animals. In other words, they are different species evolving to fill the same ecological niche. Such is the case when a number of different animals evolved to have crocodile-like jawbones and bodily features. They weren't all crocodiles, it's just that the crocodile had for some reason proved evolutionarily successful at that time, and evolution loves a winner. Droser likes to refer to them not as Lazarus taxa but as "Elvis taxa" in that most forms were primarily imitations. But there were a lot of them. About 30 percent of all such groups were Lazarus taxa during some point between the mid-Permian and the mid-Triassic.

One of the most crucial problems with acidification is the loss of coral. Coral reefs build up over time and offer shelter for smaller fish and other organisms. The coral reefs of the world are home to 25 percent of all marine species, yet they occupy a total area about half the size of France. Global warming and acidification have already led to increased levels of coral bleaching, which eliminates algae in reefs. Coral have a symbiotic relationship with various species of algae. They provide algae a place to live, and algae provide coral with vital nutrients. But coral bleaching eliminates the algae, and as a result coral starve.

About two-thirds of coral species live in deep, cold reefs, far outnumbering the more famous shallow, near-shore habitats of the Indian and Pacific Oceans and the Caribbean Sea, which are better known to vacationing snorkelers. Like shallow reefs, deep coral reefs provide shelter to an enormous and colorful bouquet of sea life. Fish that live in both deep, cold coral reefs and shallow, warm reefs represent a quarter of the annual marine catch in Asia, and feed about a billion people.

There are effects in the Southern Ocean encircling Antarctica as well. Acidification there dissolves the shells of sea snails. Geraint Tarling with the British Antarctic Survey in Cambridge captured free-swimming sea snails called pteropods and found that under an electron microscope they showed signs of strong corrosion. Experiments have shown that coral and mollusks use calcium carbonate in the water to make their shells. But increasing levels of ocean acidification means there is more carbonic acid in the water and this attacks shell building.

Certain types of phytoplankton, which have calcium carbonate shells, may be devastated by acidification in our oceans. Plankton that live in reef communities will suffer a double whammy, since acidification will destroy corals and raise temperatures above reef animal tolerances.

What happens then? Well, considering that atmospheric oxygen comes from two major sources—the tropical rain forests and marine plants such as kelp, algal plankton, and phytoplankton—deforestation and acidification may literally be attacking the air we breathe.

With ocean acidification, we may be harming the environment less purposefully than by overfishing, but the two combined are a bombshell to our present-day marine environments. It's amazing to consider, but it hasn't been that long since we started taking fish from the ocean. Archaeologists studying fish bones at 127 archaeological sites across England found a remarkable change in catches starting around 1050. According to Callum Roberts, a professor of marine conservation at the University of York, England, and author of *The Unnatural History of the Sea*, it was only in the beginning of the last millennium that people who were used to eating freshwater fish and freshwater/ocean migrants (such as salmon) began eating fish primarily from the sea.

Fish from rivers and ponds, such as pike, trout, and perch, as well as migratory fish like salmon, smelt, and sea trout dominated archaeological sites from the seventh to the tenth centuries, but from the eleventh century onward the fish bones in English digs changed to mostly herring, cod, whiting, and haddock—all sea-based creatures. New fishing technologies as well as bigger boats stoked the fishing fires, but the truth was there simply weren't enough inland fish left to feed the growing British population.

Trawling, the dragging of nets across the seafloor, goes back to the late fourteenth century. It's a destructive type of fishing that indiscriminately catches fish both big and small. Trawling nets are, however, a boon to ocean fishing.

Hook-and-line fishing enjoyed a boost in the eighteenth century when long lines with hundreds of thousands of hooks replaced hand lines with much fewer. But the true dawn of industrial fishing began in the mid-1870s when the steam trawler appeared. The fishing power of sailing trawlers had been limited by tides and wind, but the steam trawlers were forever freed from the constraints of weather. Steam trawlers quickly replaced sail power for bottom trawling. The development of the frozen food industry during the 1920s provided the next big boost.

Even so, coming out of World War II in the 1940s and 1950s, such environmentalists as Rachel Carson, author of *Silent Spring*, couldn't fathom a future without fish. Most marine experts thought the oceans were inexhaustible. They were wrong.

In the decades that followed, intensive fishing became an enormous worldwide industry. Bigger boats, longer lines, and ever-larger trawls worked the sea with an efficiency not previously possible. Doctors started talking about how fish was much better than beef for one's health, providing another big boost for the fishing industry. Global fish catches reached a peak at about 85 million metric tons a year in the 1980s. Large catches were maintained by a growing fleet with more advanced equipment.

Peter Ward, a paleontologist at the University of Washington, claims that by some estimates every square mile of the world's continental shelves is trawled every two years. But as the continental shelves have begun to diminish, fishermen have entered the last great wilderness: the deep sea. Muddy bottoms cover much of the deep-sea floor. But here and there seamounts (underwater mountains) thrust their peaks up just shy of the surface and allow for pockets of enormous fish diversity. Giant circular currents move up and down, bathing the tops of the seamounts in phytoplankton.

In the late 1960s, Soviet fishermen discovered plentiful schools of armorhead fish around seamounts off Hawaii and began to harvest them. Fish around seamounts had to contend with stronger open-ocean currents, so they were more muscular and tastier than coastal fishes. Other countries followed the Russian lead, and seamounts off Hawaii were fished intensively. But the run didn't last. Around 1976, catches collapsed from 30,000 tons to just 3,500 tons. If the Hawaiian fish bonanza had proved to be short-lived, no matter: there were plenty of seamounts left in the sea.

The next jackpot came from Soviet ships fishing at depths of 2,600 to 3,300 feet (800 to 1,000 meters) over the Chatham Rise off New Zealand in the early 1980s. Here, fishermen ran into plentiful populations of a bright-orange fish—what scientists referred to as *Hoplostethus atlanticus*, a relatively large deep-sea fish and a member of the slimehead family. But "slimehead" didn't sound like something housewives would want to unload their wallets for, so they changed the name to "orange roughy." It is still used worldwide for breaded fillets, fish cakes, and fish sticks, along with other white fish.

Fishermen in New Zealand and Australia quickly joined the Russians in a full-scale assault on the fishery. One Australian fisherman, Allan Barnett, struck it rich at St. Helen's Hill off the edge of the Tasmanian island shelf in 1989. In the first year, the hunt brought in a whopping seventeen thousand tons of orange roughy. But catches soon began to plummet as fishermen worked one seamount after another. Orange roughy are a very long-lived fish that do not reach reproductive maturity for over twenty years, making them extremely susceptible to overfishing and very slow to recover.

But that's not the end. I visited Craig R. McClain, assistant director of science at the National Evolutionary Synthesis Center, in

Durham, North Carolina. He is a husky, young, friendly evolutionary marine biologist whose specialties are deep-sea species and very large marine animals like giant squid.

According to McClain, "We have overfished the shallow seas and are now moving into the deeper waters and doing the same." He claimed the next big pressure in the deep sea is going to come from industrial mining companies that want to harvest the rare minerals in the bottom of the ocean. Mining companies off Papua New Guinea are starting to explore deep-sea vents, as they are made of a lot of precious minerals needed to make things like computers and, ironically, Toyota Prius hybrids. China is now considering harvesting deep-sea sediments for their rare earth metals.

The runoff of nitrogen and phosphorus fertilizers from inland farms that travels downriver to the oceans is another part of our marine problems. During my visit McClain told me, "We are doing the equivalent of fertilizing a forest and that completely changes the makeup of sea plants and animals. And we are warming the oceans as we are acidifying the waters. This is all radically altering the temperature and chemistry of the sea." He claimed we are suffering from severe reductions of shark species, some migrating species, and all the top predators. McClain sees the problem as reaching beyond just the loss of the deep sea, and said: "We are in danger of losing the entire ocean."

This is changing the way biologists approach the ocean. In late summer, while visiting Frank Hurd, the science director of Olazul in La Paz, Mexico, I noticed all the divers were wearing long-sleeve T-shirts and long tights, what they called full-body exposure suits, though the waters were quite warm. I kidded him about finding the waters off Baja cold. "I don't wear these things to keep warm," said Hurd. "I wear them to protect myself from the jellyfish."

According to Hurd, there have been massive jellyfish blooms along the coastlines of the US and Mexico in recent years. One of the hot spots for jellyfish is Monterey Canyon, in the center of Monterey Bay, the largest submarine canyon along the coast of North America.

The National Science Foundation, in a special report titled "Jellyfish Gone Wild," claims that one-third of the total weight of all life in Monterey Bay consists of jellyfish and similar gelatinous creatures. This is also a prime area for Humboldt squid.

Jellyfish move in as fish move out. They are tolerant of both low-oxygen environments and ocean acidification. A look into the future of the ocean could take us to the Republic of Palau, a group of islands about 550 miles east of the Philippines. About ninety thousand tourists visit Palau annually and one of their favorite haunts is Jellyfish Lake (Ongem'l Tketau to the locals), which is easily accessible by boat from Koror, Palau's capital. There are five landlocked marine lakes on Palau and each has different species of jellyfish. Scientists at the Coral Reef Research Foundation on Palau believe that the spotted jellyfish was the original ancestor of all Palau's landlocked jellyfish, but they followed the rules of evolution and morphed over time into unique species for each of the five lakes, much like Darwin's finches did in the Galápagos.

Jellyfish Lake's jellies range from the size of a blueberry to the size of a cantaloupe. Tourist snorkelers love to swim through the millions of jellies as they pulse in and out. These jellies migrate across the lake each day. They go eastward in the morning to the edge of the shadow cast by the mangrove trees that surround the lake and then reverse their course, congregating by the western shadow line by midafternoon. Though these jellies have stingers, they target crustaceans about the size of a bee in the lake. Your skin might tingle if you touched one, but it wouldn't sting.

However, the sting of a box jellyfish, found off Indonesia and Australia, can kill a man or a woman in just three minutes. Jellyfish sting about 500,000 people each year in the Chesapeake Bay, the largest estuary in the United States, on the Atlantic Coast, surrounded by Maryland and Virginia, but the US has nothing as lethal as a box jellyfish. Box jellyfish kill twenty to forty people each year.

During jellyfish blooms, about 500 million refrigerator-sized Nomura's jellyfish float in the Sea of Japan. They can grow up to 6.5 feet

(2 meters) wide and weigh 450 pounds (220 kilograms). Though normally more common off China or Korea, they have been showing up in Japanese waters, where they clog fishing lines and can poison fish catches with their toxic stingers.

A study by Cathy Lucas, a marine biologist at the University of Southampton, UK, predicts that jellyfish concentrations are regulated by decadal fluctuations and that the rise in the 2000s is part of the normal ups and downs that last peaked in the 1970s. But another study, conducted by researchers at the Institut de Recherche Pour le Développement in France, speculates that overfishing is the cause. Researchers at the Institut compared two major ecosystems along the Benguela Current, which flows along the southwestern coast of Africa. Off Namibia, where commercial fishing regulations are lax, jellyfish are spreading in coastal waters. But off South Africa about 600 miles (1,000 kilometers) south, where fishing has been tightly controlled for sixty years, jellyfish populations are stable.

José Luis Acuña, at the Universidad de Oviedo in Spain, studies jellyfish, which he says are an increasing problem in some parts of the Mediterranean. He claims that despite the fact that jellyfish are slow-moving, drifting animals with no vision to help them spot prey, they do as well as some sighted, fast-moving fish, when you factor in their much lower metabolisms that don't require as much food, and their large body sizes, which are achieved by the addition of a good amount of water.

Jellyfish are ancient, dating back 600 million to 700 million years or more. That's three times the age of the first dinosaurs. Acuña speculates that jellyfish have survived and will continue to thrive in the future by evolving large, water-filled bodies that can come in contact with more prey. Although larger bodies are less efficient, collecting your prey while drifting through the water beats the high-energy costs of hunting it down. He sides with those who think that overfishing is promoting the presence of jellyfish in ocean waters. Without eyes, jellyfish seem to be able to handle human polluted environments better than fish.

Jellyfish are adaptable, perhaps even more than humans. Our strategy has been to go after food, metals, and fuel with full and furious fervor, whereas jellyfish have evolved a much more passive strategy of gently moving though the water, taking only what they need, and limiting their expenditure of energy. We've moved in ways that are exhausting our available resources, while jellyfish glide through the water, carefully limiting their costs. Which has the best outlook for the future? In a structured competition, it would be hard to think we could outlast the jellyfish.

Is the future of the ocean one filled with jellyfish and squid? Perhaps. But both of these creatures bear the markings of "weedy species"—those that rush in after a catastrophe. They are like the new grasses that sprouted up near Mount St. Helens in places where the volcanic eruption had blown the trees down. Or the early Triassic scallop *Claraia* that moved in after the Permian extinction. They are quick to take over disturbed areas and to go through large population explosions, but those explosions are not made to last.

If the pressure from man continues at current rates, we will indeed trash the oceans. The world's fisheries are already at an extreme point. At a recent symposium for the American Association for the Advancement of Science, I heard one European scientist say that we should eat fish only on major holidays "like Americans now do with turkey."

Despite the best efforts of environmental groups like the World Wildlife Federation, the future does not look bright as long as man is in the picture. However, if we were to take a sabbatical or perhaps an early retirement, the ocean would return in due time. And we're not talking about returning to conditions as they were before the first European explorers. To resurrect life as it was in its prime, we need to go back much further.

Says Olazul's science director Frank Hurd, "Most fisheries managers try to rebuild ecosystems to what they saw fifty years ago. But if you want to know what virgin nature really looked like in the Americas, you don't go back fifty years, or even five hundred years, you go back before the first human ever appeared."

What would that look like? Man has altered the oceans so thoroughly that it's hard to imagine. Callum Roberts in *The Unnatural History of the Sea* offers up Captain Edmund Fanning's first visit to Palmyra Atoll in the Pacific Ocean in 1798 as his best attempt to describe it. Captain Fanning was headed from the Juan Fernández Islands, off the coast of Chile, to Canton, China, with a boat full of fur seal pelts. Fanning and his men from Stonington, Connecticut, had spent four months off Chile taking fur seals for their pelts. Palmyra, in the middle of the Pacific, was the halfway point of their journey.

Late on a hot June night, Fanning's men, who'd sighted breakers ahead and worried there were hidden obstacles under the water, awoke the captain. What the men had seen was an atoll of islands circling a bay wreathed in foam from the ocean swells, which were exploding against the coral beneath. Fanning and his crew struggled to find calm waters on the downwind side of the atoll, where they dropped anchor. They awoke the next morning to encounter about fifty islets surrounding three lagoons. The shores were fringed by coconut palms and coconuts lay all over the beaches, untouched by man.

Fanning and a few of his crew took a rowboat to investigate. While rowing into the bay, he was astounded at the abundance of fish. Ravenous sharks grabbed at the rowboat's rudder and oars, "leaving thereon many marks of their sharp teeth and powerful jaws." As they entered the bay, the sharks were replaced by multitudes of fish that were less rapacious but even more plentiful.

The men went ashore for coconuts while Fanning stayed with the rowboat to fish. He stood there with a harpoon and caught fifty mullet weighing about five to twelve pounds (two to five kilograms) in short order. He stopped, perhaps thinking the fish might spoil or the boat might sink under the weight of his men and the fish.

Today, Palmyra Atoll, 1,000 miles south of Hawaii, has passed through various international controls before the Nature Conservancy bought it in 2000. Though there is a small private-use airport run by the conservancy, the island is mostly the same place that Fanning visited in 1798. Coral reefs that have grown on the rim of an ancient submerged volcano form the atoll. These vast submerged reefs support three times the number of corals found in the Caribbean and Hawaii.

Palmyra is one of the few places in the ocean where top predators dominate the underwater community. Dive into the water and sharks surround you, a site not seen often elsewhere. Palmyra has more apex predators—large fish like groupers, jacks, and sharks—than any other reef known to science. A diver stepping into this unique ecosystem is stepping back in time to when fishing had not yet affected the seas. The reefs support a complex web of life: not only sharks but pods of dolphins, manta rays, sea turtles, and thousands of tropical fish.

Do you want to peek at the ocean after man? Go visit Palmyra Atoll. The species may change going into the future, but everything else will probably be the same. A rise in the sea level might submerge the reef, but the reef will eventually return once mankind hits the road.

Prior to Fanning's visit to Palmyra Atoll, there was no evidence of human contact. Fanning may not have appreciated what he was seeing, but it was something that today is extraordinarily unique.

Perhaps the greatest treasure of this magical place is its abundance of marine predators, which are currently under assault almost everywhere else in the world.

11

PREDATORS WILL SCRAMBLE

OVER THE LAST 600 million years, during most of the extinction events, predators were the last to go. During the Cretaceous extinction, which knocked out the dinosaurs, the asteroid's impact created clouds of gases and dust that blocked out light. This killed off the plants, which knocked off the plant eaters, and took out the predators, which ate plant eaters for lunch. Plants, plant eaters, and predators were the sequence then, but this time we're attacking both ends. We're killing off our plants, which are at the bottom of the food chain or "web," as biologists prefer to call it, while at the same time going after predators, at the top of the web. We're killing off predators first, either because they have valuable appendages (shark fins, rhino horns, elephant tusks) or because they take our domestic animals, or simply because, once in a while, they get one of us.

There are unique consequences for this top-down approach, says Jim Estes, professor of ecology and evolutionary biology at the University of California, Santa Cruz, when I visited with him at the Long Marine Laboratory on campus. He witnessed some of the top-down consequences in 1970 when, as a graduate student, he was sent to the Aleutian Islands between Alaska and Russia to study sea otters. One of his professors had urged him to address the role of sea otters as predators within the Aleutian ecosystem. "It never dawned on me that that would be an interesting question," says Estes.

The Aleutian Islands are a chain of volcanic islands that stretches from the Alaska Peninsula toward the Kamchatka Peninsula, creating the boundary between the Bering Sea and the North Pacific Ocean. This is an area of stormy seas where one is not likely to find cruise ships, rustic inns, or tourists. Amchitka Island within the chain was used as an airfield during World War II but is currently uninhabited. Northern sea otters were nearly wiped out by fur hunters here in the late nineteenth century but an international treaty in 1911 stopped the pillage. By the 1970s the northern sea otter had recovered over vast areas of its former range, but not all. This gave Estes a unique vantage point from which to understand the value of a predator within a maritime ecosystem by viewing islands with and without otters.

In the first weeks of the study, Estes piloted a boat around Amchitka, past submerged rocks and into foggy inlets, here and there diving under the icy waters to get a glimpse of what lay below. Around the craggy underwater shorelines, the seas were filled with kelp plants that grew up from the bottom offering a respite, nursery, and feeding grounds for a wealth of marine creatures. Kelp is one of the fastest-growing plants on earth: under ideal conditions it's capable of growing up to two feet in a day and can reach 175 feet in height in a matter of months. Beneath the surface, the kelp rises like an undersea forest. Large golden leaves attach to long thin stalks that sway with the movement of the currents.

Amchitka offered a robust, healthy ecosystem with predators in place, but Estes needed a comparative view. So he traveled to Shemya Island a couple of hundred miles west of Amchitka. Shemya had come under the same human assault that had wiped out the sea otters at Amchitka, but the otters hadn't yet returned to Shemya. When Estes entered the water there, he found a different world from the one he'd found at Amchitka. To start, there was little or no kelp. Instead, he viewed an ocean bottom thick with urchins, small spiny, spherical creatures, the favorite food of otters. The role of the otter as predator was immediately obvious. With otters, there were still some urchins present but they occupied hidden crevices and weren't numerous enough to curtail the growth of kelp. Without otters, there was a thick covering of urchins and no kelp forest. Since then, Estes has spent a good deal of his professional career trying to understand that relationship.

Without otters, there was simply no predatory pressure on the urchins, and without this pressure, urchin populations boomed. The problem was that urchins in sufficient numbers attacked the kelp holdfasts at the base of the plants, and this killed the kelp plants and the marine forests they created.

Estes visited New Zealand to understand what kept urchins at bay in those waters where there were no otters and never had been. The biologist found that southern kelp had developed a load of noxious compounds to make them unsavory to urchins. Off the Aleutian Islands, Alaska, and western Canada, the otter discouraged the presence of urchins all by itself. And with otters protected, the balance returned. Otters recovered to 75 percent of their original range in the Aleutians, as the kelp forests there grew thick and healthy.

But then, in the beginning of the 1990s, sea otter populations plummeted there once again. Their numbers, estimated at 55,000 to 100,000 in the 1980s, dropped to 6,000 by the year 2000. Some marine biologists blamed disease, others blamed increased ocean temperatures from climate change, and still others pointed their fingers at industrial fishing. But one day in 1991, Brian Hatfield, a US Geological Survey biologist working with Estes, came into one of their field offices in the Aleutians and said he thought he'd just seen a killer whale take a sea otter, but he wasn't sure. "He came back a few days later," says Estes, "and this time he was positive."

Estes didn't bite at first. Pollution, industrial fishing, and disease were still the favored culprits of the otter decline back then, but all of these options would have left a weakened, scrawny population of otters, and the otters in the Aleutians were fit and fat. Pollution, industrial fishing, and disease would have weakened the otters through sickness and poor health. But killer whales reduced the population of otters to such low levels that food was no longer a limiting factor, which was why these small furry creatures were so big and healthy.

Otters weren't the only marine mammals in trouble in that region. Populations of Steller sea lions, northern fur seals, and smaller harbor seals were also collapsing, and their remaining populations appeared healthy as well. The case for the killer whales grew stronger.

Alan Springer, a researcher as the University of Alaska Fairbanks, approached Estes at a conference and showed him how killer whales depended on large whales as a food source, but post–World War II industrial whaling had removed half a million great whales—the killer whales' natural prey—from the North Pacific. Prior to whaling, the North Pacific and the southern Bering Sea had an estimated 30 million tons of whales, but when the International Whaling Commission imposed a moratorium on whaling in 1985, only 3 million tons of whales survived. About 90 percent of whales in the North Pacific had been destroyed, and the killer whale was frantically trying to make up for the loss of food.

In 2011, Estes—along with John Terborgh, professor of biology at Duke University, and twenty-two other biologists from around the world—published an article in *Science* about how large apex predators had been a critical player in the global environment for millions of years, but that their loss might be man's most extensive and appalling legacy on earth.

The loss of top or apex predators alters the intensity of plant eaters, and this has enormous effects on the abundance and composition of plants. As wolves have returned to Yellowstone National Park and started to prey on elk, willow and aspen trees, which the elk formerly grazed in excess, have also returned.

The opposite picture has occurred on the small islands of Lake Guri, Venezuela, but it proves a point about how predators can change the color of the forest. A hydroelectric dam created Lake Guri, and as the waters rose, it formed an isolated group of islands within the lake. The area surrounding the islands had once been dense green tropical forest dominated by top predators like jaguars and harpy eagles. But as the water rose, the predators fled the islands and the forest on those islands began to change.

The Duke professor John Terborgh, a lanky, rugged ecologist who ran the school's tropical research center in the jungles of Peru for more than twenty years, noticed how, in the absence of predators, prey populations exploded. On one island, iguanas were living at ten times their normal densities. On another, howler monkeys were fifty times denser than on the mainland. On a different island, leaf-cutting ants were living at one hundred times their normal numbers. Only the toughest plants with thorns and lethal chemicals could survive the resultant assault. The island forests were sparse and brown compared to the mainland forest, which was lush and green. Terborgh suggested that predators played a major part in making the world green by controlling the plant eaters. Uncontrolled, the plant eaters turned the forest brown.

A different sort of example was the Scottish island of Rum, where wolves have been absent now for 250 to 500 years. Rum provides a peek at the consequences of predator loss, which can result in elevated browsing by deer and other herbivores. Though designated a National Nature Reserve in 1957, during its long period of predator absence, the Isle of Rum has transitioned from a forested environment to a treeless landscape.

Sharks are another vulnerable predator, but their problems are directly linked to man, not killer whales. Boris Worm, professor at Dalhousie University, in Halifax, Nova Scotia, Canada, claims in a recent paper that man kills 100 million or more sharks every year. Sharks have persisted for at least 400 million years and are one of the oldest vertebrate groups on the planet, but their populations are disappearing rapidly.

The problem is a global boom in shark fishing largely generated by an increasing demand for shark fins, used in shark fin soup in Asia.

Formerly consumed by Chinese emperors, shark fin soup is similar to champagne: something offered to celebrate good fortunes at weddings, graduations, and business lunches. However, it's a ritual that is threatening the existence of an animal that has been with us in various forms since the Devonian period. An estimated 38 million sharks are taken to feed the worldwide fin trade each year.

Fisheries kill one in fifteen sharks every year. Sharks are similar to whales and humans in that they mature late in life and have few offspring, which makes their populations uniquely vulnerable.

Peter Klimley, a marine biologist at the University of California at Davis, has been studying hammerhead shark populations on the Espíritu Santo seamount off Baja California for several decades. Klimley believes that hammerheads use Espíritu Santo as their mating grounds. Hammerhead sharks circle around the top of the underwater mount in a large school as females compete for dominant positions in the center.

Espíritu Santo's sharks don't feed while schooling there but travel to nearby feeding areas at night to gorge on squid. Klimley thinks they follow cracks in the seafloor filled with magnetic lava that radiate like spokes from the seamount. Sharks use special sense organs, the ampullae of Lorenzini, which are electroreceptors that can read magnetism like a compass. The hammerhead shark is currently on the International Union for Conservation of Nature (IUCN) endangered species list, and Klimley sees fewer and fewer sharks in the Gulf of California each year. He is currently tracking juvenile hammerhead sharks to find out if they frequent other similar seamounts, with the idea that some of those open-ocean areas might warrant protected status.

Still, hammerhead sharks don't warrant the greatest fishing pressure. That priority is reserved for great white, bull, and tiger sharks, all known for making serious, unprovoked attacks on humans. Fishermen pursue them for their meat and their fins as well as for vengeance for occasional attacks on humans.

All three sharks inhabit their own unique environments and generally don't cross paths with one another or with killer whales, though fishermen like to contemplate such encounters. Tiger sharks are common throughout the tropical world. According to George Burgess, director of the Florida Program for Shark Research and curator of the International Shark Attack File, "Tigers are apex predators throughout most of their range. Killer whales do overlap a bit on the edges of their ranges; however, I am unaware of any recorded killer-tiger interactions. In my opinion, tigers (and whites) are co-apex predators with killers in the areas they coinhabit. All three consume fishes and cetaceans and don't eat each other."

Large tiger sharks can grow to 20 to 25 feet (6 to 7.5 meters) in length and weigh more than 1,900 pounds (900 kilograms). Many of the prey that tiger sharks go after are defensive animals like puffer fish, stingrays, and triggerfish. These fish have adapted spines, teeth, and even poisons to ward off predators like tiger sharks. But according to Kim Holland, an associate researcher at the Hawai'i Institute of Marine Biology, "The tiger shark has apparently decided, 'Heck, we'll just eat them anyway.' I can't tell you how many tiger sharks' mouths I've investigated that were filled with stingray barbs."

The only other true man-eater is the bull shark. Though not as well-known as the tiger and the great white, bull sharks can be nearly as dangerous. Florida is the shark attack capital of the United States, and bull sharks attack more Floridians than any other shark species. Bull sharks swim in tropical and subtropical waters around the world.

Bull sharks get their name from their short, blunt snouts and pugnacious dispositions as well as their tendency to head-butt their prey before attacking. They are medium-size sharks that can grow up to 11.5 feet (3.4 meters) and weigh up to 500 pounds (230 kilograms).

Bull sharks are the only large sharks that can survive in freshwater. Female bulls enter estuaries, bays, harbors, lagoons, and river mouths to bear their young, which spend their early years in these habitats. Bull sharks have been sighted 2,220 miles up the Amazon River near Iquitos, Peru. They have also been reported up the Mississippi River as far north as the state of Illinois.

Humans kill about 45 million sharks a year, whereas sharks kill fewer than 4.5 of us each year, though that doesn't include fatal attacks reported as drownings. Burgess thinks shark attacks on humans are highly exaggerated, but that the damage humans do to sharks is very real. Says Burgess, "Their numbers are down by 90 percent or more in some populations."

Still, sharks are one of nature's great success stories. Some 2,000 to 3,000 species of fossil sharks have been described as compared to 650 to 800 species of dinosaurs. We still have 1,100 species of them today. Not all sharks are in danger—mainly the big ones, particularly those that eat man.

The great white shark ranges worldwide but most commonly inhabits the coastal waters of North America, South Africa, and Australia. It can measure up to 20 feet (6 meters) in length and 5,000 pounds (2,268 kilograms) in weight. As strange as it may seem, great whites have become an ecotourist attraction for caged divers off the coast of Australia.

Their role as such could provide a better future for the great white shark. A study by University of British Columbia researchers states that shark ecotourism currently generates more than $314 million a year around the world and that figure is projected to double in the next twenty years. This compares to $780 million for landed sharks, a business that is shrinking. University of British Columbia researchers examined shark fisheries and shark ecotourism data from seventy sites in forty-five countries. Chris Lowe, director of the Sharklab at California State University, Long Beach, says that the ecotourism business in Australia is an effective deterrent to shark fin fishermen. "They report anyone going after white sharks," Lowe observed. "No one wants to see those tourist dollars disappear."

The great white shark is listed on the IUCN Red List as "vulnerable." Relatively little is known about its biology. It is fairly uncommon, most frequently reported off South Africa, Australia, and California. World catches of great white sharks from all causes are difficult to estimate. The animal matures late and has few offspring, so that if populations were to suffer, it would be slow to rebound.

Off California, great white sharks are ambush predators concealing themselves in the rocky bottoms of offshore islands. These islands are home to a number of seals and sea lions, but the favorite of the great white shark appears to be the elephant seal, the largest of California's seals.

THE ELEPHANT IN THE ROOM

By attaching tracking devices to elephant seals on the Channel Islands off California, biologists have recorded that females travel nearly twelve thousand miles annually, and males more than thirteen thousand miles—the longest migration of any mammal on earth. But these animals do not just swim straight to feeding grounds; they dive continuously throughout their journeys, going to extreme depths to feed on deep-sea fish and squid and to avoid great white sharks. These dives add an average of five thousand vertical miles to their lengthy horizontal journeys. "They are basically on the move the whole time," says Robert DeLong, a biologist with the National Marine Mammal Laboratory in Seattle.

The great white shark is the primary predator of elephant seals in the North Pacific. They circle the Channel Islands during the mating and calving season in the winter, looking for stray seals that leave the safety of the beach to start on their long migrations.

If great white sharks were to go extinct, what could take their place as predators of elephant seals? What could assume the position of top predator off the California coastline? Mako sharks might fill the bill. They are legendary swimmers, reaching sustained speeds of 22 miles per hour (35 kilometers per hour) with bursts to over 50 miles per hour (80 kilometers per hour). Still, they are smaller than great whites. Their maximum length is 13 feet (4 meters). One of the largest mako sharks ever taken was 1,323 pounds (600 kilograms), caught off California on June 4, 2013. Mako sharks might have to feed in packs to take an elephant seal, and cooperative hunting is not typical of sharks.

Our triumphant Humboldt squid might be picking up that trait. William Gilly at Stanford University put cameras on these squid and recorded their hunting in tightly coordinated groups in the Pacific, a behavior that is usually associated with fish rather than squid. At present, it is mako sharks that are feeding on Humboldt squid, not the other way around. Still, mako sharks are frequently covered with scars from Humboldt squid. They appear as a ring of small incisions or a series of parallel scars suggestive of squid suckers, which have teeth, being dragged along the skin. Linear scars often begin with a circular mark on a shark's midsection and lead forward toward the mouth of the fish. At least Humboldt squid are putting up a fight.

Author and professor Callum Roberts reports that all large shallow and midwater predators are disappearing. Could giant and colossal squid move up from the deep sea and establish themselves in shallower waters? The National Evolutionary Synthesis Center's Craig McClain worked in the Bahamas in a place called the "Tongue of the Ocean," a deep trough in the waters offshore. He's also worked in Monterey, where a deep canyon occurs in the continental shelf. Areas like these and the deep ocean off Newfoundland are where giant squid and colossal squid live. Strandings of giant squid on Newfoundland shores are thought to be the result of warm-water incursions into deeper-water canyons, where scientists think giant squid reside. Still, if the loss of polar ice leads to a shutdown of the deep ocean currents that bring oxygen to the deep sea, then there could be strong evolutionary pressure for giant and colossal squid to migrate to shallower depths in the water column.

Giant squid are the biggest invertebrates (animals with no backbone) on earth. The largest of these elusive giants measure fifty-nine feet (eighteen meters) in length and weigh nearly one ton (nine hundred kilograms). In 2004, researchers in Japan took the first images ever of a live giant squid. And in late 2006, scientists with Japan's

National Science Museum caught and brought to the surface a live twenty-four-foot (seven-meter) female giant squid.

Giant squid, along with their cousins, the colossal squid, have the largest eyes in the animal kingdom, measuring some ten inches (twenty-five centimeters) in diameter. These massive organs allow them to detect objects in the lightless depths where most other animals are blind.

Like other squid species, they have eight arms and two longer feeding tentacles that help them bring food to their beaked mouths. Their diet likely consists of fish, shrimp, and other squid, though some suggest they might even attack and eat small whales. Scientists don't know enough about these beasts to say for sure what their range is, but giant squid carcasses have been found in all of the world's oceans.

The range of colossal squid is similarly mysterious, but early whalers found colossal squid beaks in the stomachs of sperm whales, so at least we know what eats colossal squid. The colossal squid is the largest squid in terms of mass. Their cousins, the giant squid, have suckers lined with small teeth, but colossal squid do their kin one better by having their limbs equipped with sharp claws or hooks.

These are set in a double row in the middle of each arm, preceded and followed by the more standard toothed suckers. The claws likely assist in holding and immobilizing struggling prey like grappling hooks as the animals are drawn to the parrot-like beak of the squid. The largest known colossal squid to date was captured off New Zealand and weighed 1,091 pounds (495 kilograms) and measured 33 feet (10 meters) in total length.

Perhaps if this deep-sea creature got bigger it would have a better chance of taking the place of the great white shark, but giant or colossal squid could also borrow a few advantages from their cephalopod brethren, cuttlefish and octopuses.

Killer whales take on much larger whales by hunting in packs, but giant and colossal squid appear to be loners. A little better communication skill could help them with pack hunting, which cuttlefish, closely related to squid, have mastered. Cuttlefish range in size from two inches to three feet. They look like short squid. As cuttlefish swim over yellow sand, brown sand, multicolored pebbles, or even beds of white shells, they can instantly and effortlessly change colors and textures to mimic the colors and textures of the bottom so predators above can't see them.

Cuttlefish also have a rich vocabulary of signals for hunting, reproduction, and warning. An intense striped zebra display warns other males to stay away. Roger Hanlon of the Marine Biological Laboratory at Woods Hole, Massachusetts, described in a telephone interview and an online seminar from his lab a number of elements of the cuttlefish "vocabulary" that might come in handy. These include defense against predators, communication with other cuttlefish, attracting mates, repelling or deceiving rivals, signaling alarm to others, and more.

The giant Pacific octopus is a cousin to the squid and has some of the squid's abilities, but also a brain that beats them all. Giant Pacific octopuses can grow as large as six hundred pounds, but most weigh less than one hundred pounds. They are found in the coastal waters of the North Pacific from Japan to California. Scientists believe octopuses are more intelligent than any of the fishes, though not as smart as most mammals.

At the Seattle Aquarium, volunteers tend to give only octopuses, seals, and sea otters names, as these animals exhibit the prerequisite personality. Roland C. Anderson, a former biologist at the aquarium, told me that giant Pacific octopuses held in their tanks will invert their bodies to expose their suckers like a panhandler when they want food. If you give it to them, they will swim back and forth in their tanks turning red, perhaps "the only example of an invertebrate showing emotion," he said.

Give a marine animal the body of a colossal squid, the brains of a giant Pacific octopus, and the communication ability of the cuttlefish and the result could give colossal squid top or apex predator potential. Add to these the newly discovered ability of Humboldt squid to hunt collectively, and they could take the place of a great white shark, hunting and feeding on elephant seals. The trick is to get them to come up from the depths. But elephant seals already dive to great depths to catch fish and avoid topside predators.

Nature could bypass the depth problem and go with Humboldt squid as the top ocean predator if they could get larger. They currently live to about one and a half years of age. No one has ever recorded finding a two-year-old squid. However, their growth rate is exponential at 5 percent a day. At one and a half years they can weigh up to a hundred pounds, but if they lived to two years, explains Gilly, they could weigh up to 660 pounds. If they lived up to three years, they could weigh up to two tons. "It would be pretty scary if these things figured out how to live longer than two years," he says.

If man takes all the great white, tiger, and bull sharks, nature could adapt, though who would reign as champion of the seas would be a matter for evolution and time to decide. Larger animals—both in the sea and on land—once existed on earth, and it's possible that they could return without us.

Part IV

NOW WHAT?

12

THE DECLINE AND RETURN OF MEGAFAUNA

As *Homo sapiens* came out of Africa sixty thousand to eighty thousand years ago and spread over the world, animals in their path gradually got smaller. Man was an ecological force against size. As he moved out of Africa, the large mammals he encountered disappeared. They were often the easiest to hunt and provided the most food. In Africa, large animals evolved with different hominids over millions of years, observed their tactics, and learned how to keep their distance from these lethal and tricky two-legged creatures. But animal species in Australia, New Zealand, and North America weren't as aware.

Man entered Australia about forty thousand years ago, and the continent lost over 85 percent of its large mammals within a period of only about five thousand years. According to UC Berkeley's Charles Marshall, "That's about a one percent change every fifty years. So if you are twenty and you look at the world around you and then you are seventy and you look again, do you notice the one percent difference? Probably not. Even the extinction of large animals in Australia took five thousand years, which, though geologically fast, was still creeping on human time scales. But we're not creeping anymore. People who do work in tropical rain forests, they go back five years later, and some of these areas are totally gone."

New Zealand had an assortment of very different creatures before man. Writes Jared Diamond in *The Third Chimpanzee*, "The scene was as close as we shall ever get to what we might see if we could reach another fertile planet on which life had evolved." The most successful of New Zealand's animals was the giant moa, an ostrich-like bird that stood over ten feet tall and weighed over five hundred pounds. New Zealand had different species of moa, the giant being the largest, instead of bison. It had songbirds and bats instead of mice, and huge eagles instead of leopards.

Polynesian settlers first landed in New Zealand only about a thousand years ago, and within a few centuries managed to annihilate the local fauna. Approximately 50 percent of New Zealand species disappeared, including all the large birds and most of the flightless ones. Moas were exploited for food and their skins and bones. Their eggshells served as water containers. The remains of approximately a half million giant moas ended up in archaeological sites, many times more than would have been alive at any one time. The settlers apparently hunted giant moas for generations before they went extinct, as man almost did in North America with the buffalo.

The extermination of many of the large animals in North America occurred from thirteen thousand to ten thousand years ago. Some scientists speculate that a sudden cold reversal of temperature that hit earth from eleven thousand to ten thousand years ago, known as the Younger Dryas event, caught millions of species unprepared. Others think it was the quickly rising temperature at the end of this event that killed off the furry creatures.

Still others believe it was the Clovis people whose fossilized bones and other relics became dominant in archaeological excavations dated about thirteen thousand years ago. Of course, there are earlier sites with evidence of human habitation in both North America and South America before this time, but this period is notable because it is when Clovis populations gathered en masse and their fluted stone arrow points, a characteristic tool, started showing up frequently.

Some believe that the demise of megafauna was itself the cause of sudden cold reversal. It may have cut off a large, important source of the strong greenhouse gas methane: the four-chambered stomachs of the animals themselves, which expelled the gas by burping (not passing gas). In other words, the early hunters shut off all the methane that the large animals were releasing to the environment, and this itself caused the Younger Dryas onset of cold by temporarily shutting off the greenhouse gas that was making things warmer.

Blaire Van Valkenburgh, a vertebrate paleontologist at the University of California, Los Angeles, agrees that humans weren't the sole cause of all the extinctions, "but they were an additional force that acted upon an ecosystem that had been in balance, but with the arrival of the Clovis people quickly became out of balance," she told me when I visited her office.

When Clovis man appeared, he was a new carnivore that was competing with the saber- and scimitar-toothed cats, which now had to share their meals with a skillful new predator. Groups of large cats preyed on bison and horses, though there is also evidence of them following mammoth herds and attacking young mammoths. Man was essentially an uninvited guest that the cats had to deal with, and there wasn't enough food to go around. Humans were the tipping point. After his arrival, North America's large animal populations started to disappear.

Van Valkenburgh believes that further proof of the competition between man and megafauna is found in the condition of the teeth of saber- and scimitar-toothed cats and the large dire wolves. These predators show heavier tooth wear and more broken teeth than modern-day beasts like cougars and gray wolves, though some of that was present even before humans arrived. Increased competition with *Homo sapiens* may have aggravated an already competitive situation.

"Large predators tend to exhibit heavier tooth wear and greater numbers of broken teeth when they consume carcasses more completely, actively feeding on bones," Van Valkenburgh and coauthor William J. Ripple wrote in a paper in *BioScience*. "The predators, which were much more abundant than the humans, most likely killed the vast majority of the megafauna." The addition of the new human predator was more the straw that broke the camel's back rather than the lead cause of the extinction.

How did these mammals get so big in the first place? While the dinosaurs were still alive, mammals scurried around under the feet of their much larger neighbors, occasionally taking refuge under bushes, in tree hollows, or in underground burrows. But once the asteroid fell from the sky and the dinosaurs disappeared, mammals started growing, and growing, and growing.

It began about 65 million years ago and peaked about 30 million years later. Animals grew about eight orders of magnitude (× 10, × 10, × 10 . . . eight times) rather quickly, but it took them about 30 million years to max out. The maximum of seventeen to eighteen tons for a land mammal has remained constant over time, in different places, and with different species, says Felisa A. Smith, a professor of biology at the University of New Mexico. *Indricotherium*, a hornless rhino-like herbivore that weighed approximately seventeen tons and stood eighteen feet at the shoulders, lived in Eurasia about 34 million years ago. It was the largest land mammal that ever existed. *Indricotherium* would have towered over a modern African elephant.

The colder the climate gets, the larger the animals, since they conserve heat better. Xiaoming Wang from the Natural History Museum of Los Angeles County and Qiang Li of the Chinese Academy of Sciences recently uncovered a large woolly rhino in the foothills of the Himalayas in the southwestern Tibetan Plateau. The animal stood perhaps six feet tall and was twelve to fourteen feet long. It had two great horns, one sprouting about three feet long from the tip of its nose, while the other arose from between its eyes. The Tibetan woolly rhino was stocky like today's rhino but had long, thick hair. It is one of the giant mammals like woolly mammoths, giant sloths, and saber-toothed cats that became extinct. It is thought to be about 3.7 million years old. Over a million years older than the previous oldest woolly rhino fossil ever found.

The Tibetan woolly rhino lived at a much warmer time when northern continents were free of the massive fields of ice that came later with the Ice Age. But, residing in the Tibetan Plateau, this animal grew accustomed to the cold at high elevations and was "pre-adapted" to it. Thus, when the Ice Age arrived, these cold-tolerant rhinos simply descended from the highlands and began to spread throughout Eurasia. The Tibetan Plateau might have served as breeding grounds for these and other giant Ice Age mammals.

FAIR TO MIDDENS

Professor Smith, whom I meet in her lab at the University of New Mexico and who towers over me in her cowboy boots, has been researching size in animals for many years and believes size is one of nature's most important adaptations to climate change.

Smith currently studies the size of present and ancient large pack rats (also known as wood rats) at Death Valley National Park in California, details of which she discovers by studying their middens, the refuse tossed out of a pack rat nest. She uses this evidence as clues to their environment and ecology. She can tell the size of the animal and indirect information about the climate by examining the size of the animal pellets or feces in a pack rat's midden. She also gets teeth and bones from these middens, which are used to confirm the identity of the species constructing the nest. She showed me some samples of these middens in her lab and even invited me to pick one up, which I did, cautiously. She asked what I thought of the smell, and I told her it smelled sweet, to which she responded, "You're a born pack rat midden researcher!" since that smell was actually pack rat urine, which the animal uses to hold the pieces together.

According to Smith, the relationship of body mass and temperature has proven so predictable it's known as Bergmann's rule (named for Carl Bergmann, a German biologist): For broadly distributed mammal groups, the larger-size species are found in colder environments and the smaller sizes are found in warmer places.

Smith studies ancient pack rat middens because they provide detailed fossil evidence of the times when they were created. Pack rats are collectors of twigs, leaves, small rocks, fecal pellets, and anything they find and deposit in the large piles of debris in front of their nests. The middens provide protection from predators and insulation from climate swings. When they are constructed on rocky outcrops, they can last for thousands of years and can be carbon-dated. A single mountain may contain dozens of middens spanning thirty thousand years or more.

The size of the fecal pellets is an indication of pack rat size and diet. Researchers are able to characterize body and genetic responses to climate in populations of pack rats spread out over thousands of years.

Today, Death Valley holds the record for the hottest and driest place on earth, but during the last ice age, Death Valley was covered by Lake Manly, a hundred miles long and six hundred feet deep. The climate was 11 to 18 degrees Fahrenheit (6 to 10 degrees Celsius) cooler. But as the valley began to warm, pack rats adapted and slowly moved upslope. They got as high as 5,900 feet (1,800 meters) but it wasn't high enough. By about six thousand years ago there were no more large pack rats present on the east side of Death Valley.

THE MEAT TRAP

Not all mammal extinctions were due to warming or man. UCLA's Van Valkenburgh argues that over the last 50 million years, successive groups of large cat-like, wolf-like, and hyena-like mammalian carnivores diversified but then declined and went extinct. At one time the Canidae family (wolf-like carnivores) had three subfamilies. Two of those went extinct. Van Valkenburgh believes that some of this was caused by what she calls "the meat trap." In situations where carnivores need more energy, they may switch not only to a pure carnivore diet but also to a diet in which their prey are bigger than they are. Once there, however, they have trouble going back to smaller portions.

All three subfamilies of Canidae reached a peak about 30 million years ago, but only one subfamily survived. That included domestic dogs, wolves, foxes, and coyotes. The other two subfamilies increased in body size by 400 to 600 percent, but when their prey got sparse, they couldn't switch back to smaller prey. They had adapted to eating all meat, all of the time, and only from animals larger than themselves. Their diets were simply unsustainable. Still, there were once twenty-five contemporaneous species of canids native to North America as opposed to seven today. Nature has indeed been much richer before. But can it ever return to the diversity it once had? Can we turn back the clock?

RETURN OF THE CALL OF THE WILD

For years, the golden era to which environmentalists in North America have often spoken of turning back the clock was before 1492, when the Europeans arrived. However, in the journal *Nature* in 2005, Josh Dolan, a biologist at Cornell University, and a group of prominent scientists expressed their desire to go back even further. They wanted to go back to a time before the Clovis people, the real starting point for human change in the Americas.

Dolan's group declared that Western scientists had gone into full retreat from the battle to stop biodiversity loss and were now simply struggling to diminish the rate. The team wanted to change the game from just "managing the extinction" to actively "restoring ecological and evolutionary processes." Their idea was to restore all the big animals that once stocked North America with surrogates from other continents that could push back the time line to when there were horses, camels, elephants, and even lions stalking the land.

One of the first things the group proposed was to restore the largest tortoise in America, the Bolson tortoise, to areas of the Southwestern United States. I accompanied the late David Morafka, a herpetologist at California State University, Dominguez Hills, to visit the Mapimí Biosphere Reserve, dedicated to the protection of the Bolson tortoise and other unique flora and fauna of the Bolsón de Mapimí, a large inland basin in the Chihuahuan Desert north of Mexico City. The Bolson tortoise, whose range had once extended across the Chihuahuan Desert in northern Mexico and the southern extremes of the US, was making a last stand here. Repatriating the Bolson tortoise to the broader expanse of the Chihuahuan desert tortoise could bring the largest of the continent's tortoise species back to the US.

Wild horses were another possibility for rewilding. They were introduced by Europeans to North America about five hundred years ago and have since taken up ecological niches that were held more in balance thirteen thousand years ago. Many ranchers look at wild horses and burros as large pests that foul watering holes and compete with cattle, native pronghorn antelope, and native bighorn sheep. To Dolan, the horse is just as native to this land as any other species.

The problem is you can't just reintroduce a large herbivore like a horse and not reintroduce predators to keep the animals under control. In 1971, the Wild Free-Roaming Horses and Burros Act made it illegal for anyone to harass, capture, or kill wild horses. Since then, populations of the animals have soared throughout the Great Basin Desert, which lies mostly in Nevada and extends to the fringes of surrounding states. But this is not the case in Montgomery Pass Wild Horse Territory, which straddles the California-Nevada border. There scientists study mountain lions, which are an effective predator control for these animals.

Such a balanced approach could provide a lifeline for Przewalski's horse, smaller than most domesticated horses and native to the steppes, the vast semiarid grass-covered plains of Central Asia, as well as the Asiatic wild ass, both free-roaming and critically endangered equids. Translocation to the US might save them from extinction and repatriate horses to their evolutionary home ground. Many scientists say that the trick is to introduce predators to these ranges as well, in order to keep large horse populations in check.

Believe it or not, camels originated in North America. They migrated north from the Arizona desert and crossed the Bering Strait land bridge three to four million years ago. The IUCN currently lists the Bactrian (two-humped) camel as *critically endangered*. There are about 600 Bactrian camels surviving in the wild in China and 450 in Mongolia. It is the only truly wild camel. Dromedary (one-humped) camels, but for some feral animals in Australia, exist only as domestic animals. There were four species of camels and llamas in North America at the end of the last ice age. Today wild Bactrian camels are restricted to the Gobi Desert and their cousins, the llamas, to South America.

In the 1850s, Lt. Edward Beale led the US Camel Corps, mostly dromedaries, from Texas to California, and he was amazed at how camels grazed on creosote and other brush species that now form dense monocultures across much of the Southwest desert. If we brought Bactrian camels back again or released domestic camels into the wild, the landscape of plants could be more diverse. Camels were once a vital part of the ecological community. Australia has well-managed co-grazing programs of cattle and dromedary camels, which could provide meat and milk, and an increase in the mosaic of plant species.

Elephants would be another winner in the west. At Olduvai Gorge, I saw elephants feeding on shrub forests. Despite a plethora of sharp spines and thorns, the elephants cleared away the brush as efficiently as tractors but left the refuse to regenerate the soil. Much of the open grasslands of East African plains owe their existence to the assistance of elephants. Introducing elephants to the juniper forests on the Edwards Plateau in Texas might alleviate the juniper problem.

Another animal that the "rewilders" would like to reintroduce to the American West is the cheetah. Cheetahs were once here. The American cheetah first appeared perhaps 2.5 million years ago but went extinct about two thousand years ago with the rest of the megafauna. They are the reason that the American pronghorn antelope is so fast. Pronghorn can travel at a top speed of sixty miles an hour, second only to the cheetah. But whereas the cheetah is a mere sprinter, the pronghorn is an endurance runner. Pronghorn can average forty miles an hour for half an hour or more, galloping across the high prairies of Wyoming. There is no living reason for pronghorn to move that fast. Biologists believe that they evolved to outrun the American cheetah, which was around when pronghorn, mammoths, and giant sloths roamed the North American plains. Returning the cheetah to America could give the pronghorn an impetus to stay fit. Also, the African cheetah, once found throughout Africa and southwestern Asia, has been greatly reduced and is not likely to survive into the next century. Moving them to the North American plains might increase their chances. This wouldn't be another case of introducing an invasive to North America, as all these species were present long before man showed up. In many ways, we're the worst of the invasives.

There are currently about a thousand African cheetahs in zoo populations across the world that could act as surrogates for the American cheetah, which is closely related. I watched a cheetah from a safari car in Ngorongoro Crater in Tanzania: the animal was grooming itself in the sun. Behind the cheetah were several antelope, something that attracted the cheetah's attention as well. The cheetah slowly got up and started to stalk an antelope, when a pair of hyenas jumped up, drawn by the cheetah's hunting pose. They bared their enormous teeth like laughing clowns, waiting for the action to begin.

The cheetah took a runner's pose and bolted, leaving a huge cloud of dust behind it. The galloping cheetah quickly caught up with the antelope, knocking it over, while the hyenas danced about, wildly excited over the promise of shared meat.

Such scenes would be great for tourists. Ecotourism has that potential even in North America to raise substantial funds that could benefit the parks that protect these animals as wells as the surrounding communities. About 1.5 million people annually visit the San

Diego Zoo's Wild Animal Park in California to see large animals. By contrast, only twelve US national parks receive that many visitors.

Rewilding might fill the excitement gap of public and private parks in the US. The reintroduction of wolves to Yellowstone National Park has generated an additional $6 million to $9 million annually at costs of $500,000 to $900,000 to the park. If the chance of seeing a wolf in the wild has generated that much support, how much support might come from the chance to see a cheetah or an elephant?

The California condor thrived throughout North America until the end of the last ice age. They roamed over the Grand Canyon ten thousand years ago, scavenging on large animals that are now extinct. Today, California condors once again soar over the Grand Canyon, but the captive breeding program that promotes this must feed these animals with cattle carcasses. If Pleistocene megafauna were returned to the American West, these scavengers might again flourish.

Today, deer populations in the Northeastern woods of the United States are at historically high levels; so are disease-bearing pests, as we've seen with black-legged ticks. The presence of disease is associated with ticks, white-footed mice, and white-tailed deer. Gray wolves once caused deer to avoid heavily wooded areas, where they were more vulnerable to attack. But without gray wolves, white-tailed deer frequent the wooded forests, where ticks and Lyme disease reach their highest incidence. If wolves were to return, a lessening of disease risk, including Lyme disease, hantavirus, monkey pox, typhus, bubonic plague, and hemorrhagic fever might result from a better-balanced ecosystem.

These megafauna proxies proposed for rewilding are not exactly the same species that existed at the time of the Pleistocene extinction, but they could fulfill similar roles, as did the reintroduced birds in the North American peregrine falcon program. That program attempted to restore to North America peregrine falcons, whose eggs had become too brittle for hatching due to the use of DDT, a commonly used pesticide that was banned in 1972 but persisted in the environment for years afterward. The program used large numbers of captive-bred birds of different subspecies to bolster American falcon populations. In the end, these birds adapted to fulfill the niche left by the Midwestern peregrine population, which disappeared in the 1960s.

Biologists with cautious controls would carefully monitor a Pleistocene rewilding program while staying as true to the fossil record as possible. Private lands would hold the most immediate potential. More than seventy-seven thousand large Asian and African mammals now occupy Texas ranches. Larger tracts of public lands in the Southwestern United States could be brought on board to expand the program.

Bolson tortoises and exotic species of horses might be the first logical step, since they so recently occupied similar lands in North America. Camels and llamas might follow, since these animals could help control invasive plants. The final introductions might be elephants and African lions. The benefit of having elephants and camels for the control of woody vegetation has already been explained. But the benefit of African lions is more controversial. The African lion was once the widest-ranging land mammal of all time. The Asiatic lion is critically endangered, with a single population in India's Gujarat State. Yet lions have been introduced and managed in African and Indian reserves that are a similar size to some contiguous and private lands in the US.

Establishing a predator population would be a necessity. The central issue, of course, is that lions sometimes attack humans. Such a reality has been growing in acceptance with mountain lions in the US. Attacks don't precipitate large kills of lions anymore. But African lions would be an upgrade in size and predator status. The African lion is the apex predator of Africa.

The African and Indian reserves that have reintroduced lions have been successful in reestablishing normal behavior and population controls in their prey. But momentous questions would have to be answered before the reintroduction of cheetahs and lions could begin in the US.

Wolves were introduced in the 1990s to Yellowstone National Park and they have contained burgeoning elk populations there sufficiently to allow a resurgence of the forest. They have also started to take coyotes, not just as prey, but as competitors, which reduces the coyotes' take of pronghorn antelope fawns and other smaller predators, like raccoons and beavers. Wolves also reduce or scare off hoofed mammals that trample streamside vegetation, and this enhances nesting habitat for migrant birds.

Before man migrated to North America, there were many more predators and prey in considerably grander and better-balanced ecosystems. The situation that exists now is diminished, impoverished, and unnatural. We have fewer birds, animals, reptiles, and amphibians—fewer of almost everything. "But we are incredibly adaptive, which means we don't remember the past well," says UC Berkeley paleobiologist Charles Marshall. "Where I grew up in Australia, the area was just packed with wildlife and all sorts of natural noises. In comparison America is much more impoverished. It has far fewer natural sounds. But I've adapted. I don't notice the relatively silent days anymore. But I sure noticed it when I got here."

Scientists tell us that Southeast Asian jungles used to be raucous forests with abundant wild noises. But these jungles have become the key supplier of the international wildlife market, providing animals for food, traditional medicine, pets, trophies, and decorations. The demand is fueled by the region's economic growth, increased personal wealth, and the rising popularity of traditional Chinese medicine, both in China and abroad.

According to Liz Bennett, vice president of species conservation at the Wildlife Conservation Society in New York, poaching is pervasive across Southeast Asia, creating "silent forests," empty of wildlife, throughout the region. She experienced this most starkly at Kubah National Park in Malaysia, on the island of Borneo. "It's a beautiful forest, all the trees are intact, there are insect noises, but you don't hear gibbons in the morning, you don't hear birds singing, you don't even see squirrels."

Rewilding might be an exotic concept to Americans, but not to the Dutch. Flevoland, a province of central Netherlands, formerly resided at the bottom of an inlet of the North Sea. An enormous drainage project lifted Flevoland from the muck of its former seafloor in the 1950s. Today, Flevoland houses Oostvaardersplassen, a wilderness erected from what was formerly mud. Biologists have now stocked these fifteen thousand acres (six thousand hectares) with animal types that would have occupied the mainland and Flevoland—if it hadn't been underwater. Most of the original animals have gone extinct, so biologists looked for surrogates. Instead of aurochs, large and now extinct bovines, they brought in Heck cattle, red deer, and Konik horses (a primitive breed native to Poland), which live in wild herds in a natural manner. They also added white-tailed eagles, ravens, foxes, egrets, geese, and other creatures. Now herds of large animals roam over this Dutch park, which looks from a distance like the Serengeti. Visitors pay up to forty-five dollars to take a safari-like tour of the place.

Such is the success of the Dutch experiment that it has inspired Southern Europe's own rewilding movement. Every year thousands of acres of marginal farmland is taken out of production, some of it to counter climate change, and those have been suggested as future parks. The Fundación Naturaleza y Hombre (Nature and Man Foundation) recently released a herd of twenty-four Retuerta horses, one of the oldest horse breeds in Europe, in the Reserva Biológica Campanarios de Azaba in Spain. With large-scale land abandonment now happening in many parts of Europe, it provides an opportunity for nature. The creation of this new population of Retuerta horses will help to guarantee the survival of this rare breed in an area with black vultures and black storks. Rewilders are also working on the Reserva da Faia Brava in Portugal, which hopes to mix Portuguese horses with Bonelli's eagles, golden eagles, griffon vultures, Egyptian vultures, and eagle owls.

* * * *

The idea behind rewilding is noble, but the question is: Will the sixth mass extinction leave the world intact, or will man bulldoze or burn the last of earth's open spaces before he turns the light off? The causes of mass extinctions past—whether they were volcanoes, asteroids, or man—have left devastated landscapes from which it took a long while to recover. If rewilding leaves no legacy and man should exit stage left, then where will the fauna and floral of the future emerge from?

University of Washington paleontologist Peter Ward writes in *Future Evolution: An Illuminated History of Life to Come* that domestic and urban species are the best candidates for a future laid waste by man. He believes that domestic animals and plants will be the dominant members of what he calls the recovery fauna and flora (animals and plants), those species that follow a mass extinction. He claims that domestic species are already taking the functional place of extinct or endangered life.

Domesticated plants would have problems, says Scott Carroll, the director of the Institute for Contemporary Evolution, in Davis, California. "They are really like pets. We're basically watering, fertilizing, and tending them like house plants." A few adaptive plants like date palms might persist and grow as weeds, but he's not optimistic about corn or any of the major domestic grains.

The most successful domestic animal species display rapid maturity, the ability to breed in captivity, a reluctance to panic when startled, and, perhaps most of all, an amenable disposition. If domestic species exhibited these traits, they survived. If they didn't, man killed them. For this reason, domestic animals are basically dumber than their ancestors. Dogs are dumber than wolves, cats are dumber than lions, and cattle are dumber than bison. Carroll thinks domestic horses might persist. "Horses are bred to run, which would give them a chance against wolves and mountain lions. Cattle, on the other hand, are not so good at running," he says.

Sure, these animals have taken the place of the megafauna on the grasslands of the world, but could sheep, goats, or cattle exist without human guardians? It's hard to imagine domestic animals taking over the wild—or whatever wild that man leaves intact.

To understand what might have been the ultimate wild, I visited the La Brea Tar Pits, which surround the Page Museum on Wilshire Boulevard. On an overcast day in Los Angeles, Caitlin Brown and Mairin Balisi, two graduate students at UCLA, took me on a tour of the museum and its famous tar pits. We visited Pit 91, the centerpiece of one of the richest pockets of Ice Age fossils in the world. The walls of the pit have been shored up with railroad ties, and the entire excavation is surrounded with glass. The pit is an actual tar seep from underground oil deposits of the Salt Lake Oil Field about a thousand feet below the surface of a section of Los Angeles called Hancock Park. Pit 91 is seven miles west of downtown Los Angeles and three miles south of Hollywood.

Early Los Angelenos came here in the 1800s to collect asphalt for road building and other purposes. Early bones found here were dismissed as domestic animals until 1875, when William Denton, a geologist, visited the tar pits and identified the canine tooth of a saber-toothed cat. Still, people didn't realize what a treasure the pits were until 1901, when another geologist identified the bones as having come from extinct species. A paleontological gold rush ensued until, panicked at the prospect that the place would be overrun, George Allan Hancock, the owner of the land, granted the Natural History Museum of Los Angeles County exclusive rights to excavate the pits from 1913 to 1915.

The museum unearthed nearly a million bones from approximately a hundred sites at the pits. Those bones included saber-toothed cats, dire wolves, American lions, and short-faced bears—all bigger than their modern-day equivalents. The North American giant short-faced bear weighed up to 2,500 pounds (1,134 kilograms). Its cousin, the South American prehistoric giant short-faced bear, tipped the scales at up to 3,500 pounds (1,600 kilograms), perhaps the largest carnivore of its day. The pit also gave up the remains of camels, ground sloths, and mastodons.

During the summer months, the asphalt at the surface turned into a thick, sticky mess, quickly acquiring a deceptive surface of dust and leaves, making these pits an efficient animal trap. If a ground sloth, camel, or mammoth ventured out on the surface, it would take only about an inch or two of tar to totally entrap it, leaving the animal open to starvation and predator attack.

In winter, the asphalt would turn hard again, slowly interring the bones captured in the warmer seasons. One of the most interesting findings at the La Brea Tar Pits is that carnivores outnumber herbivores by almost nine to one. Even birds caught in the tar were predominantly birds of prey. The noise of the mired herbivores may have served as bait for predators that then also became mired. Once these bones were covered with tar, they endured little weathering, as tar is a good preservative.

Because pits would open at different times, they presented unique windows into our world from twenty-seven thousand to forty thousand years ago. The George C. Page Museum was built at the La Brea Tar Pits in 1977 and the bone collection transferred to it. Pit 91 was reopened and is currently under excavation. The museum staff, UCLA academics, and assorted volunteers are currently emphasizing smaller prey, carnivores, birds, and plants in the hope of understanding the ecosystem that existed back when the giant animals roamed the earth.

I met with UCLA paleontologist Blaire Van Valkenburgh, who heads up the school's contribution to the excavation, at her office later that spring day. She claimed that the La Brea Tar Pits provide an accurate window into life and the ecosystems surrounding it prior to the Pleistocene extinctions and the rise of man. Her research has been focused on large predatory animals. Modern species evolved with larger and much more complex predators, and one of her goals is to understand how those animals affected the ones today.

The Page Museum has a mechanical statue of a saber-toothed cat on the back of a giant ground sloth, which was one of nature's largest beasts in the Pleistocene. The mechanical cat sinks its enormous canines into the neck of the ground sloth over and over, but this is a mischaracterization. "In reality, a saber-toothed cat wouldn't have been able to bite into the animal's back," Van Valkenburgh pointed out to me. "Ground sloths are related to armadillos and have small nodules of bone, which formed in the skin, under the fur. Saber-toothed cats wouldn't have been able to penetrate that. The cat would have had to go for the neck of the sloth, and the ground sloth could have crushed the saber-toothed cat in its arms. Ground sloths were very powerful."

A more typical encounter at the pits thirty-six thousand years ago would have involved a pack of dire wolves quickly dispatching a camel. They might be challenged by large condors circling overhead while large coyotes hovered nearby, eager to feast on the scraps, but only the saber-toothed tiger, which had twice the mass of the wolves, could have moved the wolves off their kill. Says Van Valkenburgh, "By the following day, little evidence would have been left of the camel's death, and over time their bones would have been entombed in the tar. It is the fossils of just such scenarios that we study today."

According to Van Valkenburgh, the biggest change between life thirty-six thousand years ago and today is the diversity of large animals that inhabited North America until the end of the last ice age, about ten thousand years ago. About thirty-six thousand years ago there were fifty-six species of hoofed mammals the size of a wild pig or larger. Today, there are only eleven. Fifteen carnivores the size of a coyote or larger preyed upon mastodons, mammoths, bison, horses, and camels. Now only the coyote comes near this part of Los Angeles.

In the Pleistocene, this city was a wide floodplain with lush vegetation and multiple streams and rivers rushing down from the tall mountains that surround the basin. The climate was cooler, wetter, and greener than it is today, more like the Big Sur area, which is three hundred miles north and famously forested. These were coastal lands that attracted migratory animals that followed the coastline with the seasons, feeding on the plants and animals along the way.

This scenario, then, is what life would have been like without us: a paradise of green lands and wild creatures and birds next to blue oceans equally rich in fish, whales, and marine animals. Though the view afforded by the scientists who work at the La Brea Tar Pits is a bit gruesome, it was not always that way.

Multiple predators ganging up on single prey wasn't the norm. In fact, scientists speculate that only one of those scenarios in a decade would have been enough to produce the fossils found at La Brea. There were plenty of green plants, shrubs, and trees around. Predators in Los Angeles thirty-six thousand years ago would have kept the animal herds stronger and the vegetation greener.

This area now surrounded by tall buildings used to be a wilderness paradise filled with creatures the likes of which we can hardly imagine. But the only way it could presently return to that idyllic scenario is without man.

So what are the chances? Could man survive a mass extinction? Is there an escape—a way out?

13

INVADERS *TO* MARS?

IF WE SPOIL THE EARTH, should we try another planet? Interplanetary travel could be a major force for human change. Other planets could have different atmospheres, dissimilar amounts of cosmic radiation, varying periods of night and day, wildly divergent temperatures, and drastically disparate amounts of gravity. All of these are strong evolutionary forces that could over time change man into something quite different from what he is here on Earth. The change in gravity alone could make the difference. In discussions about such travel, Mars is often mentioned.

Giovanni Schiaparelli, director of the Brera Astronomical Observatory in Milan, Italy, first seriously investigated Mars in the late 1800s when, through the observatory's telescope, he counted more than sixty crisscross marks on the face of the planet. Schiaparelli referred to them by the names of famous earthbound rivers, identifying the markings as a system of *canali*—the Italian word for channels. But it was the descriptions by Percival Lowell, a wealthy New Englander, author, and astronomer, whose book *Mars and Its Canals* (1906) identified those canals as a planet-wide system of irrigation—the work of intelligent inhabitants of a dying Mars, who had constructed them to utilize water from the polar ice caps—which really fired the public's imagination. Lowell's astronomy showed that colonizing Mars might not be the answer, since somebody was already up there.

On a cold December night, as the temperature dove near 20 degrees Fahrenheit (minus 7 degrees Celsius), I ventured into the observatory that Lowell built for himself in the small city of Flagstaff, Arizona. As a guest of the observatory, I got to look through the twenty-four-inch (sixty-one-centimeter) lens on the giant refractor telescope that the firm of Alvan Clark & Sons built for Lowell in 1896. Mars was in the heavens as the attendant swung the telescope about the rounded dome of the observatory. It was a couple of hours past sundown as I gazed through the fine instrument, trying to find the so-called *canali* Lowell had everybody excited about.

It was difficult to make out any canals. The planet's craters earlier had been clear and sharp, but the canals looked iffy. Particularly when you knew that modern high-resolution mapping of the Martian surface by spacecraft showed no such features. Many felt the canals were an optical illusion, perhaps what one would see after staring through the telescope long enough and late enough. But the Panama and Suez Canals were built in the decades leading up to Lowell's discoveries, and perhaps he felt that everyone must have been building canals.

Nevertheless, he and his assistants spent more than a decade mapping a system of hundreds of canals on the surface of Mars. His telescope was one of the finest made in its day, but had nowhere near the resolution that today's telescopes offer. An infrared spectrogram of Mars taken at the Mount Wilson Observatory near Los Angeles in the 1960s revealed that the Red Planet, as Mars was called for its visible reddish hue, had extremely low atmospheric pressure. The pressure on Mars was about 4.5 millimeters of mercury, compared to 760 millimeters of mercury on Earth. At 4.5 millimeters of pressure, water acts like dry ice. At its melting point it changes directly from a solid to a gaseous state. It made Percival Lowell's idea of a system of canals on Mars impossible. Water couldn't flow on the surface of Mars.

Yet photographs of the surface of Mars, taken by the Mariner orbital mission as early as 1971–72, as well as photos taken by the Viking missions, show that the planet's ancient surfaces are marked with branching networks of valleys that clearly resemble river- and streambeds on Earth. In 2008 the *Phoenix* landed successfully on the north pole of Mars and found pure water ice. Was Lowell at least partially correct?

The fourth planet from the sun is more like Earth than any other body in our solar system. The surface of Mars is more rugged, as it is older and less subject to repair. Mars has dry-ice fields, craters, volcanoes, floodplains, canyons, chasms, and tall mountains. Olympus Mons stands about 16 miles (25 kilometers) above the Martian surface and covers an area 374 miles (624 kilometers) in diameter, about the size of the state of Arizona. The Valles Marineris is over 1,850 miles (3,000 kilometers) long and covers about one-fifth of the circumference of the planet. The Grand Canyon on Earth is only about 500 miles (800 kilometers) long.

Mars has long attracted the attention of stargazers, since it is often considered the closest place we could run to if life grew inhospitable on Earth. It could also be a jumping-off station to the mineral-rich asteroids that orbit nearby. With its low gravity, Mars might prove to be a springboard to distant stars in our galaxy. *Mars Odyssey*, in orbit around Mars since 2001, used an infrared camera and a gamma ray spectrometer to map the content of the Martian surface, finding large regions near the poles where the soil had over 60 percent water ice by weight.

Scientists believe these watery observations are proof that Mars once had a warm, wet atmosphere that was suitable for life. Early Mars had a lot more CO_2 in its atmosphere than it does today, and that produced a considerable greenhouse gas effect and a much milder climate. These conditions persisted on Mars about four billion years ago, close to the point where life evolved on Earth. Could life have evolved on Mars about the same time? Is life on other planets a possibility? With so many millions of stars and millions of planets around them, how could we be the only one?

A day on Mars is similar to one on Earth, being twenty-four hours and thirty-seven minutes long. Mars rotates on its axis and has four seasons, but since the Martian year is about 669 days, winter, spring, summer, and fall are about twice as long as those seasons on Earth. The present-day Martian environment would be a little rough for humans without a good space suit. Daytime temperatures on Mars can get up to 63 degrees Fahrenheit (17 degrees Celsius), but at night they dive down to minus 130 degrees Fahrenheit (minus 90 degrees Celsius). It would be an inhospitable place for a moonlit walk, despite the fact that Mars has two moons.

The question of whether there was ever life on Mars has been an ongoing puzzle for scientists. In 1976, the National Aeronautics and Space Administration (NASA) sent *Viking 1* to the Chryse Plains on Mars with a few experiments NASA hoped would answer the life question. The Gas Exchange Experiment was set up to douse Martian soil with "chicken soup," a nutrient-rich solution that, when added to the soil, might make something breathe. On July 20, *Viking 1* set down and extended an arm out of the landing craft, scooped up some soil, and gave it some soup. And as soon as the soil tasted the concoction, there was a violent eruption of oxygen.

But other experiments didn't yield similar results. Scientists speculated that the surface of Mars might be covered with "superoxides" formed by intense UV radiation that had bombarded the surface. These superoxides had reacted to the water in the soup and gave off oxygen. "It's the same as when you pour hydrogen peroxide on a cut," says Christopher McKay, a planetary scientist at NASA's Ames Research Center. "It fizzles and eats up all the organics present."

A more recent exploration by the *Curiosity* rover in 2013 analyzed a powdered sample of soil and found some promise. Only a half mile from the landing spot in the middle of the three-mile- (five-kilometer-) high Gale Crater, *Curiosity* sampled a rock that contained sulfur, nitrogen, hydrogen, oxygen, phosphorus, and carbon—a sampling of the major ingredients of life on planet Earth.

In 2011 the *Mars Reconnaissance Orbiter* found seasonal streaks that formed and disappeared on a Martian slope and may have been the result of underground water ice that thawed and flowed in the Martian spring. Much of Mars's water is held in permafrost soils or ice. Robert

Zubrin, author (with Richard Wagner) of *The Case for Mars: The Plan to Settle the Red Planet and Why We Must*, says, "Current knowledge indicates that if Mars were smooth and all its ice and permafrost melted into liquid water, the entire planet would be covered with an ocean over 100 meters deep."

Mars may once have had a warm and wet climate suitable for the origins of life. In their first billion years or so, both Mars and Earth had carbon dioxide atmospheres and were covered with water.

We know that life evolved on our planet, but did it evolve on Mars? Is life a million-to-one long shot that could hardly occur anywhere else, or is it a natural occurrence of certain environmental conditions? If we found living organisms or simple fossils on Mars, it might mean that the universe is full of life. And that would be big indeed. It could perhaps be the escape hatch for man.

The greatest hurdle to the continued exploration and space station development on Mars is, like many things, money. Where does one get enough? When John F. Kennedy launched the Apollo program, which sent men to the moon, the US and Russia were in the middle of the Cold War, and competition and national pride were behind the big push to the moon. But the Cold War days are gone, and nothing like that has arisen to move the Mars program forward. Some say we should wait for technology to advance, to reinvent itself, but Zubrin says time is a-wasting. He feels we can get to Mars with what we've got: technologies based on Saturn V rockets from the Apollo days with engines and boosters developed during the space shuttle era.

Mars is a bit out there. At its closest orbital position, it is around 38 million miles (56 million kilometers) distant. The best time to launch a trip from Earth to Mars would be when the planets are at their maximum distance. Over the long trip, eventually the two planets would come closer together, and the trip would then be made over the smallest distance.

Then there's the problem of gas. Zubrin feels that it's too difficult to go to Mars and return home with enough gas to make the round trip. One of his most daring proposals is that we get our fuel not from Earth but from Mars. He believes that we need only carbon and oxygen from Mars and a little bit of hydrogen (about 5 percent) that we could bring from Earth. Carbon dioxide could be pulled straight out of the Martian air, which is 95 percent CO_2. Take a jar and fill it with activated carbon or other suitable material and set it out in the supercold Mars night. With a nighttime chill of minus 130 degrees Fahrenheit (minus 90 degrees Celsius), the material will soak up 20 percent of its weight in CO_2. When the sun comes back up, the material will warm up and we will generate high-carbon gas.

The idea is to send the unmanned apparatus to Mars, let it process the fuel first, and then send the manned mission when the gas station is full and in place. The first missions might have enough gas to go both ways, but the extra weight would require additional thrust, and if we can make rocket fuel from Martian air, we'll be way ahead of the game.

Once we got enough CO_2, we could mix it with the hydrogen we brought from Earth and get methane and water from the combination. The water produced can be split into oxygen, which could be stored, and hydrogen, which could be recycled back into the methane-producing process. The equipment necessary for methane production would comprise three reactors, each three feet (one meter) long and five inches (twelve centimeters) in diameter.

Scientists think that the first missions could be dangerous, and Zubrin agrees but thinks that a small crew would still be best. It would include two mechanics—or flight engineers, if you will—a biologist, and a geologist, four people in all. That would provide two scientist/mechanic teams, one at the base camp and one out in the field. The geologist would explore the planet's geological history while evaluating the planet's fuel and geological resources. The biologist could address the question of life on Mars while evaluating the soil and the environment for their ability to support greenhouse agriculture.

We could make plastics out of hydrogen and CO_2. Mars soil is full of clay, so we could make great ceramics for pottery, including pots, dishes, and cups, as well as bricks. One of the most accessible materials on Mars would be iron. It is this ore in the soil that gives Mars its reddish color. Carbon, manganese, phosphorus, and silicon are common and could be mixed with iron to make steel. Mars also has a lot of aluminum.

Silicon is plentiful, too. This could be used to make photovoltaic panels, which could generate power, though getting enough will be a problem in the early years. Though probably not a popular idea, Zubrin thinks it would be necessary to import a nuclear reactor from Earth to meet the energy demands of the base's earlier years. Once the base is well established, solar, wind, or geothermal power could be added to the mix. But nuclear power would be necessary to get things going, unless one wants to eat up the fuel needed to get home.

Geothermal power would be an attractive source, and perhaps an alternative to a nuclear reactor. Geothermal power is the fourth-largest source of power on Earth, behind combustion, hydroelectric, and nuclear. The idea is to utilize the heat of the inner planet to boil liquids and to use the steam produced to run a generator. If explorers found a geothermal heat source near underground water, that would be an inviting location for a Martian base.

Mars has other precious materials, including deuterium, the heavy isotope of hydrogen, a key element of nuclear power. There is about five times as much deuterium on Mars as there is on Earth, and a kilo of deuterium is worth about $10,000.

But perhaps the biggest attraction to building a station is the possibility of interplanetary trade. Mars is close to the main asteroid belt that circles the sun between Mars and Jupiter. Asteroids contain large amounts of high-grade metal ore, making them attractive for commerce. An average asteroid about one kilometer in diameter could hold about 200 million metric tons (10 percent larger than a US ton) of iron, 30 million metric tons of nickel, 1.5 million metric tons of cobalt, and 7,500 metric tons of platinum, worth about $150 billion for the platinum alone.

A Mars station could be a staging ground for travel to other places in the solar system and beyond. Under Zubrin's plan, the modules that house the Earth-to-Mars portion of the trip could be repurposed as the first houses of a new Martian settlement. Bricks fashioned from the finely ground, claylike dust that covers the surface of the planet could be used for additional support. These modules could be used to construct Roman-style vaults or large atriums.

Houses would have to be built underground. The Martian inhabitants would need at least 8 feet (2.5 meters) of dirt on top of their houses to properly pressurize them and to protect their inhabitants from the wide swings in temperature. Large plastic inflatable structures could be used as temporary housing while underground structures and aboveground greenhouses are being built for eventual crop growth.

Mars's atmosphere is sufficiently dense to protect its initial builders and farmers from solar flares, and there are other beneficial qualities as well. Martian sunlight, though less than that on Earth, is enough for photosynthesis. Add some CO_2 to your greenhouse and that could make up for the diminished sunlight. Martian soil is richer than that on Earth. It may need extra nitrogen, but that can be synthesized as it is here. Raising cattle, sheep, and goats would be inefficient, since it would take five times as much grain to feed the cattle as it takes to feed humans directly, so Mars astronauts might have to forgo steak in the early years.

The first Martian task would be to find water. Evidence from past missions says it's there. For manned missions, it might work to bring some more of the hydrogen (H) component from Earth to make H_2O, but once the building phase ensues and the Mars population begins to grow, water would have to take precedence. A geothermal source with water would be great. Let's just hope it's not too close to the poles. Observations by the *Mars Reconnaissance Orbiter* in 2009 reported pure water ice in relatively new craters located between 43 and 56 degrees north latitude, and that is an area of relatively temperate Martian climate.

Though the sum total of Zubrin's suggestions may sound daunting, the technological hurdles we've surmounted in just the last century make anything seem possible. There is an adventuresome spirit in man that could make it happen. Think of Captain Robert Scott and his expedition to the South Pole. Hopefully a trip to Mars might have a happier ending.

As we look to the future, Mars might also be a good place to understand our past and perhaps even the riddle of first life. Two-thirds of the surface of Mars is 3.8 billion years or older. And Mars is a lot less volcanic than Earth. Since it is small, less than twice the size of our moon, the Red Planet cooled more quickly than Earth and developed a thick, immovable crust. The surface of Earth is constantly renewed as the continental plates collide, sink, and are rebuilt, a product of plate tectonics. Earth's fossil record has yet to cough up the earliest steps that led to life, the appearance of cells, photosynthesis, and DNA. The hope is that Mars, whose crust has remained stagnant for aeons—leaving any fossils far more intact—might be a better place to understand the formation of life than here on Earth.

The draw here is that MIT and Harvard researchers think it's possible that all life on Earth is descended from microorganisms on Mars that were carried aboard meteorites that traveled to Earth. The climates on Mars and Earth were once much more similar, so life that was viable on one planet might also survive on the other. Also, an estimated one billion tons of rock have traveled from Mars to Earth, blasted loose by asteroid impacts and then hurled through interplanetary space before striking Earth's surface. And microbes have demonstrated an ability to survive the initial shock of such an impact, as well as the fortitude to journey through space and arrive on another planet.

So what about current life on Mars? Though things look a little rough on the Martian surface, could life exist underneath? Scientists have been looking at the deep and dismal corners of our planet to find out just how tough an environment life can withstand. I interviewed Bob Wharton, who passed away in 2012: he was a rugged researcher who studied karate under Chuck Norris and discovered life in Antarctica at the bottom of frozen lakes. At Lake Hoare in Taylor Valley, about eight hundred miles from the South Pole, his crew spent a half day melting a hole in the twenty-foot-thick crust of ice before climbing in and descending to the lake bottom. What they found were bizarre microbial mats—tissuey structures that are pigmented green, red, and purple to catch the limited light. "It's a fairly advanced form of life," said Wharton. "You've got a cell wall, and you've got DNA inside the cell to pass on information to its offspring. It's not elephants, but it's a big step in the evolution of biology."

Despite a mean temperature of minus 28 degrees Fahrenheit (minus 33 degrees Celsius) above the ice, underneath everything's toasty and above freezing. The ice provides what scientists call "thermal buffering." Wharton also looked for life on the 14,179-foot (4,322-meter) volcanic summit of Mount Shasta in the state of California. He sampled microorganisms there in acid hot springs. "The water would have burnt holes in my clothes," said Wharton, "but microorganisms were thriving." Could life survive in similar environments on Mars?

Even if life doesn't exist there now, it could someday. Part of making Mars a more hospitable place to live might require some monumental efforts to change the atmosphere there by a process known as terraforming: deliberately modifying features of the planet to be more like Earth. We could take a lesson from our own global warming problems and start releasing CO_2 into the atmosphere of Mars. Melting Martian polar caps would be a good start. That would liberate CO_2 and possibly methane, both greenhouse gases locked up in the permafrost. The liberated CO_2 would thicken the atmosphere and, like pulling on a blanket, would warm the place up nicely.

Zubrin has several ways to get this blanket growing. One is to establish factories on Mars to produce artificial greenhouse gases. Raising the temperature of the Martian south pole by 7 degrees Fahrenheit (4 degrees Celsius) this way could initiate a runaway greenhouse effect, which could trap even more heat. One promising and long-lasting greenhouse gas fit for the job would be halocarbons such as chlorofluorocarbons (CFCs), the kind formerly used as coolants in refrigerators and as propellants in some aerosol cans. However, we would have to choose our halocarbons carefully, picking only those without chlorine.

"Using CF gases [as opposed to CFCs] will allow the ozone to persist, while the CF gas adds to the greenhouse effect," says Zubrin. Fluorocarbons, such as CF4 (Tetrafluoromethane), could be used instead of CFC gases to create a greenhouse effect without destroying the ozone layer.

Martian explorers might use large orbital mirrors to concentrate sunlight on the poles. A space-based mirror with a radius of 77 miles (125 kilometers) reflecting light back on the Martian south pole could do the trick. An aluminized mirror about four microns (four-thousandths of a millimeter) would weigh about 220,000 US tons (200,000 metric tons), which would be impossible to haul from Earth. But the space-based manufacture of the mirror could be accomplished on a Martian moon or an asteroid.

There is a political aspect to this as well. If America wants to pick up the race to Mars again, the nation would have to time it right. The US window of opportunity, according to Zubrin, is eight years. That's the maximum length of an American presidential administration. In 1961, President Kennedy set a goal of reaching the moon by 1970. By 1968, administrations had changed, and even as the Apollo astronauts were landing on the moon, President Nixon was putting the brakes on future projects.

With the shuttle flights terminated, the US human space flight program appears to be in limbo, and space travel may have to be an international effort. In 2012, a Russian-Chinese effort called Fobos-Grunt ("Phobos-Ground") was set for an ambitious sample return mission to Mars's largest natural satellite, Phobos. However, Fobos-Grunt failed to perform an orbit-raising maneuver two and a half hours into its flight, and it never left Earth's orbit. The aim of taking soil samples on planetary moons remains a respectable one, but it may be one for the future. Remember, there were multiple failures in both the US and Soviet space programs before there were successes.

Carl Sagan, a popular American astronomer, cosmologist, and prolific author, was a strong advocate for a combined American-Soviet effort. He saw it as a way of bringing together former rivals and building trust, but both sides were reluctant to share missile technology, which could be used to send warheads. The shuttle-Mir program provided a taste of the benefits of cooperation. Between 1994 and 1998, space shuttles made a total of eleven flights to Mir, the Russian space station. American and Russian scientists also conducted experiments to determine how animals, plants, and humans would endure in space. With the demise of the US shuttle program, cooperation has diminished.

Zubrin thinks that the best choice is for the US, perhaps in conjunction with Russia, the European Union, and China, to offer a prize of, say, $20 billion for the first private organization to land a crew on Mars and to return them to Earth. This path could bring down the costs of space travel substantially. Zubrin believes that space travel under bureaucratic control is a recipe for high prices, and he thinks the private sector is often vastly more efficient because it does not require consensus to try something new. You need only one innovator and one investor.

According to Zubrin, the real cost of such a mission, pared down and under private control, would be closer to $4 to $6 billion. In this scenario a $20 billion prize would be a nice incentive. Offering a range of prizes could get things going: Let's say $500 million for a successful Mars orbiter imaging mission. Perhaps $1 billion for the first system that uses propellants of Martian origin to lift a 4.5-US-ton (4-metric-ton) payload from the surface of Mars to its orbit. And a $20 billion grand prize to someone or some organization to send at least three crew members to the Martian surface, remain there for one hundred days, take three overland trips of at least thirty-one miles (fifty kilometers), and return the crew safely to Earth.

J. Craig Venter, through his companies Synthetic Genomics and J. Craig Venter Institute, based in Maryland and California, is trying to develop a DNA sequencing machine that could be landed on the surface of Mars, look for life in the soil, sequence it, and beam it back to Earth—the benefit being that the task could be accomplished without having to return the machine to Earth. Jonathan Rothberg, founder of Ion Torrent, in Connecticut, a DNA-sequencing company, is working on a similar effort.

Mars One, a Dutch nonprofit foundation, wants to set up a permanent space colony on Mars. The company thinks that the sale of broadcasting rights of a Mars reality show would be enough to finance an actual mission sometime in 2023. The company would start with televised episodes covering the selection of astronauts, trip preparations, and the flight to Mars. After landing, the company would then start streaming operations continuously from the surface of the Red Planet.

The Mars One plan is to launch 2.75 US tons (2.5 metric tons) of supplies in 2016, a Rover in 2018, and about six landers with living pods, supplies, and support systems in 2020. The first four astronauts wouldn't arrive until 2023. A second human group would join them in 2025. The catch, however, is that the trip would essentially be one-way: there are no planned return flights. You would live your life out on Mars, and your remains would be cremated. The company says that the Martian community would decide what to do with your ashes.

Still, the company says they've had more than 100,000 applications from would-be astronauts eager to make the trip. Mars One will offer "to everyone who dreams the way the ancient explorers dreamed" the opportunity to apply for a position in a Mars One mission.

Lots to offer here, just no welcome-home party.

The greeting was more forthcoming when I pulled up to Biosphere 2 in the high desert near Tucson, Arizona. Pristine desert grasslands speckled with mesquite trees as well as prickly pear, cholla, and saguaro cacti surround the facility at the foot of the Santa Catalina

Mountains. John Adams, the assistant director of the facility, took me inside to a real-time mini world with a tropical forest, a million-gallon ocean, a small savanna grassland, a fog desert, and mangrove wetlands. Biosphere 2 is a large futuristic structure of glass atriums covering an area equivalent to 2.5 football fields. The facility was one of the first to experiment with what life might be like on another planet, though its purposes today are a bit different.

"Biosphere 2 offers a way to study the effects of climate change on these different ecosystems but in a controlled environment. Scientists here are essentially performing carefully monitored laboratory-type experiments, but on a much grander scale," says Adams.

Biosphere 2 began its own evolution as a different experiment. Its original purpose was to test the ability of man to survive in a closed, self-contained system—one completely shut off from the outside world, such as the one the explorers might encounter on Mars. The "Landscape Evolution Area" of the modern facility, now used to study soil formation, was formerly the "Agricultural System." Space Biospheres Ventures, the original developers of Biosphere 2, had hoped that food grown in it would satisfy the nutritional needs of the first eight pioneers who entered the facility in 1991.

That team was dependent on the facility's different biomes and infrastructure for the food they ate and the air they breathed—which turned out to be the things that gave project directors the most trouble. The experiment lasted two years. The first year was rough for the gourmets in the group. The crew lost an average of 16 percent of their pre-entry body weights. However, Roy Walford, a professor of medicine at the University of California, Los Angeles, and the medical doctor for the first Biosphere 2 experiment, was then promoting a low-calorie, nutrient-dense diet as a way to increase longevity. So even though the team claimed "continual hunger" in their first year of isolation, Walford happily reported that the group's cholesterol and blood pressure both went down.

But the researchers here suffered from more than loose pants; they also needed to adjust to the levels of CO_2, which fluctuated wildly.

Most of the pollinating insects died, though insect pests like cockroaches boomed. Morning glories overgrew the rain forest, blocking out other plants. The worst was that oxygen inside the facility, which began at 20 percent, fell gradually over sixteen months to 14.5 percent. The project began pumping oxygen into the system to make up for the failure, but the press caught them and cried foul.

Space Biospheres Ventures officially dissolved on June 1, 1994, after a second mission failed and federal marshals served a restraining order on the on-site management team regarding questions of authority. If Biosphere 2 had been on Mars, the occupants might have starved to death or succumbed to slow asphyxiation.

Biosphere 2 is an example of how long-term occupancy of a space station on a planet that is millions of miles from Earth could be extremely dangerous and fraught with perils that science may not yet know enough about.

On the positive side, if we can overcome these hazards, then a Mars station might offer a place where *Homo sapiens* can truly differentiate—becoming a new species. Carol Stoker, a planetary scientist at NASA's Ames Research Center, envisions a permanent research base of closed environments on Mars as the next most logical place to live outside of Earth. Still, she claims a child who grew up on the Red Planet, with one-third the gravity of Earth, would never have the physical or skeletal structure to survive on our Blue Planet.

"It is likely that a second-generation Martian would be physically unfit to walk unaided on Earth, at least without intense weight and strength training," says Stoker. "Just imagine if you suddenly weighed three times what you weigh now. Could you walk? Would your deconditioned heart be able to pump the blood volume needed? Whether we know it or not, we are constantly doing a lot of work against gravity."

The European Space Agency, the National Space Biomedical Research Institute (in Houston, Texas), and the Russian Federal Space Agency recently completed a 520-day experiment locking six "marsonauts" in a simulated spaceship near Moscow. Five hundred and twenty days is about what it would take for a round-trip flight to

Mars, with about thirty days to explore the surface. During the entire simulation, the crew went without sunlight, fresh air, or fresh food.

There were significant human problems to work through. With regard to rest, there were no external cues, such as sundown, to let the astronauts know when it was time to sleep. They had to rely on artificial cues like watches and other astronauts waking them up. Without gravity, the body had trouble telling what was up and what was down. In space, the natural orientation of the body is taken away, particularly for arms and legs. On Earth, vision, hearing, and touch combine to tell you where you are. You feel the floor under your feet, the chair that you sit upon. But weightlessness takes away those feelings, and the senses send confusing signals to the brain, which results in motion sickness.

The big thing, however, was the effect to bodily organs, particularly the cardiovascular system. While in space, the body no longer feels the downward pull of gravity that distributes the blood and body fluids to the lower extremities. Fluids start to accumulate in the upper body, away from the legs and feet. In space, astronauts actually start to look different as their faces puff out from the additional fluid in their upper bodies. They develop bird legs as the circumference of their legs shrinks due to decreased fluid in the lower body.

The heart has less work to do because it takes less energy to float around a spacecraft than it would to walk or run around planet Earth. Bones lose calcium, making them weaker, and the muscles atrophy because gravity is not providing the normal resistance to movement. Exercise machines aboard the flight can mitigate some of these effects, but that doesn't eliminate all the aftereffects. Most Russian cosmonauts, after spending months in space, were carried away from the spacecraft on special stretchers. At their homecoming receptions, climbing the podium was often too challenging so soon after reentering Earth's gravity.

Interplanetary travel would be a major evolutionary force for Earth-born settlers on Mars, and frequent travel between Earth and Mars would be unlikely because of the expense. Living on Mars could produce long-term biological changes that would make a return to Earth ultimately impossible. With isolation a natural part of the job, the gradual push of evolution toward becoming another species could happen in outer space just as well as here on Earth.

Gravity wouldn't be the only selective force. Others would include breathing compressed air and adjusting to different loads of UV radiation. The need to eat, go to the bathroom, have sex, give birth—all these vital functions would be seriously altered by changes in gravity, air, and radiation.

But even though such a change would be an interesting step in the evolution of man, it doesn't answer the primary question of life on Mars. Is it someplace where large portions of our population might escape if we mess things up down here?

There are so many things that could go wrong, one of which is that the reality TV show on Mars gets canceled due to lack of an audience and the venture runs out of cash. And there are other "little" things, like what happened with Biosphere 2's oxygen, which wasn't expected. Biosphere 2's soil was rich in organic material that was taken up by microbes, which used up oxygen and created a lot of CO_2. The plants in the facility should have been able to process the CO_2 and produce more oxygen, but it was later determined that calcium hydroxide in the concrete was removing the CO_2 and not releasing oxygen. No one would have imagined that the concrete in Biosphere 2 might eventually suffocate the residents.

After Mars, the next likely place for life in our solar system is the moons of Jupiter. That planet has four large moons and at least forty-six smaller ones. Io, for example, is the most volcanically active body orbiting Jupiter. Io's surface is covered by sulfur in different colorful forms, and its volcanoes are driven by hot silicate magma. One could keep warm next to an Io volcano, but it would be hard to get insurance. Europa's surface is mostly water ice. It may be covering an ocean of water or slushy ice beneath. This would create a "habitable zone" for microbes but it wouldn't be a place where you would want to spend your vacation, let alone the rest of your life.

Scientists are currently looking for habitable planets like Earth around other stars in our solar system, and they've found numerous prospects. But the closest star in our solar system, Alpha Centauri B, is about 4.37 light-years away from our sun. One light-year is about 6 trillion miles (10 trillion kilometers), which means Alpha Centauri B is about 26 trillion miles (41.5 trillion kilometers) from our sun. NASA's *Kepler* planet-finding spacecraft has found Earth-like planets around distant stars, only they're 275 times more distant.

Interstellar travel could happen one day in the distant future, but it's just as likely that mankind will have exhausted Earth's natural resources and made the planet unlivable before then. Right now mankind seems uninterested in either goal. What's the chance that evolution could provide the world with another species that could outcompete us and change the course of human history?

14

IS HUMAN EVOLUTION DEAD?

MANY SCIENTISTS hold the belief that natural evolution stopped for *Homo sapiens* about forty thousand to fifty thousand years ago in Europe. That's when man began to chart his own destiny apart from nature. Human inventions like sewing needles provided warm clothes to protect against the cold as opposed to natural selection providing more hair. Man began to think in symbols, which morphed into words, and this expanded into complex language. And language provided the key to elaborate cooperation.

This wasn't just hoots and loud calls with others in a group for the purpose of bringing down an animal. Language was useful for establishing trade that could reach across vast distances, relaying experiences across large chunks of time, and learning where the best food was and how to get it.

Man began to utilize an ever more complex set of tools: spears, spear throwers, bows and arrows. He grew leaner. He didn't need large muscles and thick bones to kill game at close range—he killed from a distance. The new weapons rewarded those with a better throwing arm and a better aim. The atlatl (spear thrower) and the bow allowed modern humans to kill large animals without having to have large muscles. Thus, humans could run faster, cover more ground, and not have to eat as much.

With the invention of nets, harpoons, and hooks, humans began to fish. It was less dangerous and required less physical effort. One could now eat meat, fish, and berries—a broad-based diet that gave man the advantage in a world that was then in the middle of an ice age. The use of fire and pottery for cooking made large teeth less vital and man's jaw and teeth began to recede. Cultural innovation began to affect evolution.

Yet Stephen Jay Gould, the late Harvard paleontologist, looked at all this and hesitated to give it an evolutionary consequence. Gould thought that fifty thousand to one hundred thousand years was but the blink of an eye in evolutionary development and far too rapid to see any significant evolutionary changes. But Gregory Cochran and Henry Harpending, anthropologists at the University of Utah, in Salt Lake City, say there has actually been an increase in genetic change in modern man in the last ten thousand years. In their book *The 10,000 Year Explosion: How Civilization Accelerated Human Evolution*, they propose that not only has human evolution not stopped, it has accelerated. Their belief is that evolution is now happening about one hundred times faster than the long-term average of our species' existence. Could that lead to a new species?

To arrive at this figure, Harpending and Cochran analyzed data from the International HapMap Project, an effort to describe the common patterns of genetic variation in the human genome with a goal of uncovering the genetic roots of complex diseases. The project gathered results from eleven different populations around the world, spotlighting evidence from specific sites in the human genome that influenced gene expression.

"We can compute the average amount of change in the human genome and it's one hundred times faster," says Harpending. "Which make sense. There's one hundred times as many people, and that creates one hundred times as many targets for genetic mutations."

As genes develop, so do mutations, which are genes that don't look or act as they did before. Most of these mutations are discarded in favor of the standard set, but once in a while a favorable genetic mutation occurs, with the result that people with the mutation have more children, are better adapted to fight off disease, or simply live longer. Such mutations offer an advantage: their owners do better and are more likely to survive. When this happens, the mutation is selected for and is passed on to future generations. Harpending and Cochran looked for these types of favored mutations in the human genome in regions of unshuffled genes, which indicate recent selection, since nature regularly reshuffles its genes.

These favorable genetic mutations have helped man in different ways. Man living at higher altitudes had to adapt to less oxygen in the air. To accomplish this, Andeans developed barrel chests and blood that held more oxygen, while Tibetans developed faster breathing to take in more oxygen. Scientists from the Beijing Genomics Institute recently found a set of genes in Tibetans that have helped them adapt to low oxygen levels. These "new" genes were only three thousand years old.

Harpending and Cochran also found that 7 percent of human genes underwent evolution as recently as five thousand years ago. And a lot can happen in five thousand years. Darwin chose domestic animals to illustrate much of his *On the Origin of Species.* Dogs come in enormously varied shapes and sizes. Take, for instance, a Chihuahua, which averages 7 pounds (3.2 kilograms), and a Great Dane, which averages 115 pounds (53 kilograms). Both come from the same ancestor. Neither of them looks like a wolf, yet most breeds of dog were derived from wolves in the last two hundred years.

Man is also changing. The last ten thousand years have seen numerous genetic changes to human bones and teeth along with the rapid evolution of our diet and our adaptions to disease. We are taller. Our life expectancy is much greater. Changes in society have led to evolutionary adaptations. Harpending says that we are getting less alike, so that we are not merging into a single mainstream human type. We are not the same humans we were one thousand or two thousand years ago. This may account for part of the differences between the Viking invaders and their peaceful Swedish descendants.

Harpending's coauthor, Cochran, says: "History looks more and more like a science fiction novel in which mutants repeatedly arose and displaced normal humans—sometimes quietly, by surviving starvation better, sometimes as a conquering horde. And we are those mutants."

As *Homo sapiens* migrated into Eurasia, evolution produced changes in skin color and adaptations to cold. Some of the biggest changes came with the transition to agriculture. Larger populations and more dense living conditions promoted virulent epidemic diseases like cholera, typhus, yellow fever, malaria, and smallpox. But over time this led to the development of some genetic resistance to those diseases.

Neanderthals, a species that developed in Europe, had adaptations to climate that other *Homo* species never developed in Africa. As we have discussed, resistance to such diseases as malaria is far more prevalent in Central Africans than Northern Europeans. Skin color is another important adaption to environment. Monkeys and other primates have pale skins under their fur, but humans that lost their fur, perhaps to sweat more freely, evolved darker skins to protect against ultraviolet light. The process reversed itself when man first ventured northward, where his skin grew lighter, maybe to better synthesize vitamin D.

Peter Grant at Princeton University worked on the Galápagos Islands and likes to lecture his students about the persistence of evolution, claiming that evolution is always happening. Genes of this generation are not the same as the last. Nor will they be the same in the next. He claims it's a mathematical certainty. Genes keep changing. You may not notice it. The trees around you may look the same. And the birds and the squirrels may look similar year after year. "They aren't," Grant said in an interview with author Jonathan Weiner for

The Beak of the Finch: A Story of Evolution in Our Time. "They're different. But you can't see it, the differences are too subtle."

Evolutionary changes are proceeding at a genetic level, and sometimes they are heritable and apparent—the difference in height and longevity between you and your grandparents—but most times they are not.

One of the most game-changing mutations to the human genome was lactose tolerance. It enabled man to digest milk beyond infancy. It is responsible for the largest human expansion in history, that of the Indo-European language family.

The term "Indo-European" refers to the family of related languages that spread over western Eurasia, the Americas, and Australia. It includes Spanish, English, Hindi, Portuguese, Russian, German, Marathi, French, and numerous other languages and dialects. Today there are over three billion native speakers, close to half the human population on earth.

The idea of a single large linguistic family first arose from similar observations of people from England and people from India. Sir William Jones, the chief justice of India, mentioned these similarities in a lecture in 1786, and scholars began to trace its history through linguistics and archaeology. The first or Proto-Indo-Europeans raised cattle, sheep, and goats. They were warriors, the young men gathering into brotherhoods with challenging initiation rites.

They appeared about five thousand years ago, perhaps where modern Turkey is or in the grasslands farther north. They raised stock and grew grain but depended more on animal husbandry than on farming. They expanded their dominions, it is thought, by military conquests driven by the domestication of the horse but also by the genetic mutation that gave them lactose tolerance.

Indo-Europeans originally used cattle to pull their plows and wagons, but also to provide humans with beef and leather. But as lactose tolerance spread, more people began to keep their cattle for milk rather than meat. This was a major advantage, because dairy farming is much more efficient than raising cattle for slaughter—dairy producing about five times as many calories per acre of land.

Proto-Indo-Europeans were perhaps the most competitive in areas less ideal for growing grain. It was thus easier to tend dairy cattle year-round than to try to grow grain. As dairymen, they were more mobile than grain growers, who had homes and villages to defend. Still, they had to protect their cattle, since cattle could walk and were a lot easier to steal. Early Proto-Indo-Europeans must have spent a lot of time stealing each other's cattle, and retaliating for earlier raids. Lactose tolerance produced healthier and more robust populations, though they had to fight to maintain their high standards.

Lactose tolerance also developed separately on the Arabian Peninsula, which was dependent on the domestication of camels rather than cattle. And cattle herders in East Africa also acquired it. The increase in food that dairy cattle provided produced a powerful evolutionary draw for a variety of people. This is apparent in a number of African populations including the Maasai.

LACTOSE TOLERANCE IN AFRICA

I got to witness this up close when Miriam Ollemoita, a Maasai tribal member and part of the Olduvai Vertebrate Paleontology Project, led me away from the UC Berkeley field station one summer day toward her village in the grassy plains above the Olduvai Gorge in Tanzania. It was late June, the beginning of the dry season, and we passed a middle-aged woman at a shallow well dug into the dry creek bed who was scooping water with a cup and placing each cup into five-gallon buckets while other females and their giggling children waited in line with their buckets to have the middle-aged woman fill theirs.

The Maasai are a pastoral people who live principally off milk. They are lactose tolerant, something rather rare among Africans. But they also eat dried meat on special occasions and occasionally mix blood into their milk. It is apparently a healthy diet, since most of the Maasai men at the research station were tall and lean yet agile. Leslea Hlusko, who codirected the project, used Maasai men to help her locate fossils and found them strong and able.

The village was surrounded by a living wall of twisted branches with sharp thorns and spikes harvested from nearby brush. A woman picked up and moved a bundle of brambles that acted as a gate and allowed us to pass into the village. Inside the wall, we encountered a second wall, which Miriam told me was a safeguard against lions, leopards, and cheetahs, allowing the villagers time to respond if these predators got over the first wall.

Past the second wall we ran into a group of goats herded by young boys, who gathered around to watch as two of them milked the goats. Miriam told me that much of the herding was divided into three groups, with the younger boys guarding the goats, the teenage boys guarding the sheep, and the grown-up men guarding the cattle.

The cattle had already been moved from the village to an upland area that had year-round water. Miriam told me the men, including her husband, went with the cattle. She took me through another wall to the center of the village, where a group of boys and several women guarded the newborn goats and sheep. The villagers had erected three walls to protect these animals, since lions, leopards, and cheetahs considered baby goats delicious.

Back in the 1940s, the Maasai were driven out of many of the wildlife parks, including the Serengeti. Their culture was not always compatible with national park rules, particularly since the 1980s, when tourism in these parks became the leading source of income for Kenyan and Tanzanian governments. But traditionally young Maasai men were supposed to take down a lion to be initiated into the tribe. To do this, villagers ringed the local forest and drove a lion into the path of the Maasai initiate, who had to kill the lion with a spear to prove his manhood. Some Maasai members will tell you this initiation rite is no longer practiced, but others, including Miriam, say it still is.

The village was a group of dome-shaped huts made of bent branches. A woman knelt on the top of one of the structures, spreading cow dung over the branch frames. I entered one of the huts behind Miriam and was greeted with pitch-darkness. She led me to a small bench beside a small, smoldering fire in the middle of the hut. Smoke from the fire drove out mosquitoes and other insects. As my eyes slowly adjusted to the darkness, I could see that there were little alcoves around the fire where groups of children and adults gathered. Within the alcoves, Maasai beds made with tightly stretched cattle hides were strung wall to wall. The Maasai in the hut were as friendly as they were in the field station, even allowing me to take their picture.

Though the Maasai are not directly related to Proto-Indo-Europeans, they have arrived in the present with some of the same genetic adaptations that gave Indo-Europeans control over much of the world. The Maasai may not have garnered as much of the world's material wealth, but lactose tolerance had given them a way of life that has made their tribe the strongest and noblest in the region.

Scientists believe that *Homo sapiens* was once pushed to the edge of extinction in East Africa. About seventy-five thousand years ago an eruption occurred on the Indonesian island of Sumatra. The eruption created Lake Toba, the largest crater lake in the world, but it sent three thousand cubic kilometers of rock into the air in a giant plume that spread west over Africa and Asia, enveloping everything in dust, ash, and rock. Giant rafts of pumice filled the Indian Ocean, and some of it even reached Antarctica. Dust blocked out the sun and stopped photosynthesis, which killed the vegetation and therefore the creatures that depended on that vegetation for food. Cheetahs, chimpanzees, tigers, and orangutans were pushed to the edge of extinction along with the native population.

Because of this, the numbers of *Homo sapiens* may have shrunk to several thousand, about the size of an urban high school. The evidence for this genetic bottleneck is the vast similarity between this group and modern humans. We are almost indistinguishable from each other genetically. The foreign bacteria in our intestines are more variable than the cells in our own tissues. The Lake Toba eruption is partly to blame for this lack of biodiversity.

The blast appeared around the same time as man's great cultural advancement, about the time that we started talking, painting the walls in caves, making jewelry, and conquering the world. The famous evolutionary biologist and author Richard Dawkins suggests in *The Ancestor's Tale: A Pilgrimage to the Dawn of Evolution* that the bottleneck created a situation whereby rare genes—Neanderthal DNA or some other mutation—spread through our species. Charles C. Mann, who wrote *1491: New Revelations of the Americas Before Columbus*, describes it as the moment when *Homo sapiens* 1.0 upgraded to *Homo sapiens* 2.0.

But a bottleneck limits the diversity of genes to a point whereby the species as a whole is more susceptible to a single calamity, whether by epidemic or a sudden change in climate. Our genetics, in other words, could foreshadow our extinction.

According to findings published in 2012 in the journal *Nature Communications*, large-bodied herbivorous dinosaurs were declining during the last twelve million years of the Cretaceous, while midsize herbivores and carnivorous dinosaurs were holding their own. Did sudden volcanic eruptions or an asteroid impact strike down dinosaurs during their prime? Stephen Brusatte, a Chancellor's Fellow, School of GeoSciences, University of Edinburgh, and lead author of the paper, says, "We found that it was probably much more complex than that, and maybe not the sudden catastrophe that is often portrayed." The Cretaceous extinction, which killed off the dinosaurs 65 million years ago, may not have been the "terrible weekend" scenario that some scientists like to believe. The dinosaur extinction may have been rooted in a much longer-running process that made the dinosaurs susceptible to the asteroid as well as the volcanic activity that was ongoing at that time in the Deccan Traps in west-central India, one of the largest volcanic features on earth.

Says Olazul's science director, Frank Hurd: "Eliminating so many other species of animals, lowering the biodiversity of life in general, may have been convenient for *Homo sapiens*, but in the end it may lower our own outlook for survival."

GLOBAL WARNING

Climate change has been on meteorologists' radar for several decades. Years back it was front-page news in the scientific community as well as the popular press. Now if you attend any of the science conventions, there appears to be a sense of resignation: it's happening, so we'll have to adapt to it. Man seems reluctant to make the changes necessary to stop it.

We are currently in an interglacial period where the climate has stayed rather stable. The trouble is we've come to expect it. Our present interglacial period is simply the most recent interglacial in a series of glacial cycles that have warmed and cooled the earth now for more than 2.5 million years.

IPCC predictions based on past evidence entered into computer models (to determine how climate will change in the face of rising greenhouse gases) predict that mean average global temperature will rise from 3.2 to 7.1 degrees Fahrenheit (1.8 to 4 degrees Celsius) by 2100. This is their "best estimate," from a range of estimates that go as high as 11.5 degrees Fahrenheit (6.4 degree Celsius). These predictions are partly gleaned from cores drilled into the Greenland and Antarctic ice caps as well as into the ocean floor. Some of the ice cores even bring up samples of ancient air to measure. To get a perspective on how grave those predictions are, you must consider that the difference between the current interglacial period and the last ice age is only about 10.1 degrees Fahrenheit (5.6 degrees Celsius).

Highly resolved ice cores from Greenland and Antarctica reveal twenty abrupt shifts in climate during the last ice age. In other words, abrupt climate change is part of the climate picture. We're spoiled right now because things have been so stable, but climate can shift suddenly and dramatically and remain that way for long periods.

The Younger Dryas event is one of the best-known examples of abrupt climate change. About 14,500 years ago, the earth's climate began to shift from its cold glacial world into a warmer interglacial one. Partway through this transition, however, temperatures in the northern hemisphere suddenly reversed, returning to near-glacial conditions. The Younger Dryas event is named after the Dryas flower, a cold-adapted plant common in Europe during this time. The end of the Younger Dryas, about 11,500 years ago, was particularly abrupt. In Greenland, temperatures rose 18 degrees Fahrenheit (10 degrees Celsius) in a single decade.

Man has been around for two hundred thousand years and has gone through two glacial cycles, so we may be more resilient than we're given credit for. But man has so altered the terrain of planet Earth that there is no longer enough room for nature to adapt. Species that once moved north or uphill to deal with climate change may find roads, parking lots, cities, and megastructures in the way. We've put most of our plants and animals into tightly controlled parks, so they can't leave and migrate north when the weather gets too hot.

During the last interglacial period, the Eemian, the world was a lot hotter. Ocean surfaces toward the peak of the Eemian rose six to ten feet and stayed that way for several thousand years. Salt water covered much of Northern Europe, turning Sweden and Norway into an island. Salt water also covered the western Siberian plains. The Nile River overflowed, providing a cap to Mediterranean waters that cut off the supply of oxygen to the bottom, producing thick layers of organic ooze recorded today in sediment cores taken off the coast of Egypt. Forests blanketed the Sahara and extended their ranges much farther north than they do today.

At the height of the warming, hippos pranced and snorted in the Thames River not far from the present city limits of London. Rhinos ran through the British brush, and water buffalo lowered their horns to drink from the Rhine.

* * * *

A return to this type of warming exists in permafrost soils, which underlie much of the Arctic and are large reservoirs of organic carbon—four to five times as much as all the carbon emitted from all fossil fuel combustion and human activities in the last 160 years. The permafrost has already warmed by as much as 4.5 degrees Fahrenheit (2.5 degrees Celsius). If climate change causes the Arctic to get warmer and drier, most of the carbon will be released as CO_2. If it gets warmer and wetter, most of the carbon will be released as methane. Neither scenario is cause for optimism.

Take carbon, for example: large carbon releases are already appearing across the interior of Alaska and across the North Slope. And then there is methane: in 2012 scientists measured methane releases over swamps in the Innoko Wilderness in Alaska that were similar to what one might find over a large city.

The Arctic will be one of the first areas to go. The Arctic ice rests on the sea and is only six to nine feet (two to three meters) thick, unlike the ice that covers the continent of Antarctica, which averages 7,086 feet (2,125 meters) in thickness. According to scientists, both the thickness and area of Arctic summer sea ice have declined dramatically over the last thirty years.

Summer may soon be ice-free in the Arctic, but is this a bad thing? An ice-free Arctic, after all, is a valuable potential resource for many countries. Open water in summer and thin ice in winter could be a bonanza for some of the people of Alaska, Canada, Scandinavia, and Russia, though not for wildlife. Curt Stager, author of *Deep Future*, claims an Arctic passage from Europe to the Pacific will save ships the cost of passing through the Panama Canal. An Arctic passage from Rotterdam to Seattle would be 2,000 nautical miles shorter. And a similar route bypassing the Suez Canal would cut 4,700 nautical miles off the trip from Rotterdam to Yokohama, a boon to international trade.

By some estimates, between a tenth and a third of the world's untapped oil reserves lie beneath the shallow continental shelves in the Arctic. Large volumes of gas and coal lie beneath the North Slope. Development of these petrochemical bonanzas may be an economic force for open oceans, just as petrochemical exhausts may bring them on quicker. Summer Arctic sea ice, which hit a record low in 2007, will probably dissolve completely by 2030.

But the downside is a rising ocean. Sea level has been rising steadily over the last century. Part of this is due to the thermal expansion of the ocean waters spurred on by increasing warmth, but another part is due to melting glaciers and ice caps. Mountain glaciers are melting rapidly, as are the boundaries of Greenland and Antarctica. Greenland is up to 10,500 feet (3,200 meters) above sea level, which means its high altitude supports its ice. But as the rim and base of the ice begins to melt, the peaks could lower into warmer air. It is one of the feedback effects scientists are investigating that may accelerate future changes.

The loss of Greenland's ice could raise the sea level worldwide by about twenty feet (six meters). Right now Greenland's interior ice is growing, while the margins are melting. Likewise, East Antarctica is growing while the West Antarctic peninsula is shrinking. The shrinking comes from warming seas, the growth from moisture blown over the land, which turns to ice. Increased warmth will stop the growth and increase the melting.

How could this affect modern man? It seems a lot of our megacities could become casualties of sea-level rise. London is perched on a low-lying river just upstream from a strong tidal estuary. If, as some predict, storms and floods accompany sea-level rise, then the Thames Barrier, the world's second-largest movable flood barrier, erected downstream of Central London, could suffer the same tragedy as barriers set up to protect New Orleans before Hurricane Katrina, perhaps within this century.

Amsterdam may go even before London, since it's lower. Venice and New Orleans might also have to relocate to higher ground. Large portions of Bangladesh, southern Florida, and the coastal plain of Southeast Asia, where lots of people have settled, will also be under water. Jan Zalasiewicz, a geologist at the University of Leicester, UK, and author of *The Earth After Us: What Legacy Will Humans Leave in the Rocks?* says all this is possible with a sea-level rise of twenty meters, which he claims is the "small change of geological history."

But Curt Stager argues that we don't need twenty meters to cause real damage. After only a three-foot (one-meter) rise, the Florida Keys, the Everglades, and the Mississippi Delta, along with New Orleans, will go under. So, too, will much of the San Francisco Bay Area, much of eastern China, and the southern tip of Vietnam. Also, the Dutch interior, the southwestern rim of Denmark, and the broad deltas of the Nile, Niger, Orinoco, and Amazon Rivers.

Climate change will increase rainfall in the Adirondacks in upstate New York, and this will increase downstream river discharge as well. New York Harbor will rise with the rest of the ocean around the rim of Manhattan Island, while destructive flooding coming from upstream may push the Hudson's water up and over its banks. That naturally clean New York drinking water may become a thing of the past.

Most scientists believe we are headed for a climate that is going to warm to temperatures not seen in the last ten million years in just the next few centuries. Once there, those temperatures could last over a thousand years. Back in the days of the Kyoto Protocol (an international agreement to reduce greenhouse gas emissions from industrialized nations by 5.2 percent that was adopted in Kyoto, Japan, in December 1997), there was hope we might actually tackle this, but the outlook today looks dim.

But there is another force looming as well. Have we forgotten the ice ages? The dominant weather influence of the last half million years is actually 100,000-year cycles of ice followed by 10,000-year periods of warmth. At various times in Earth's history, the planet has frozen from pole to pole with ice. Even the oceans have turned to ice. This happened once at 2.5 billion years and again between 700 million and

800 million years. Severe ice ages have occurred 400 million, 300 million, and 200 million years ago. It could happen again—though, ironically, Gifford H. Miller of the Institute of Arctic and Alpine Research at the University of Colorado, Boulder, suggests that global warming may actually put off the next ice age by thousands of years.

So we will still have our dance with global warming, but after a few thousand years of that, then comes ice. Another ice age would radically reduce the volume of plants and animals. We will have to do more with much less. Overpopulation would become lethal. What's left of us will have to rely on agricultural land that won't be as rich as it once was. But, as before, we'll be plagued by that hunter-gatherer mentality that has proven so lethal to this world: Move into a new area, use what there is, and move on. Don't worry about the future.

The future will take care of itself. Though, perhaps, not in a way we would approve of.

I met biologist Rob Jackson at his new Stanford University office one day and we discussed Anthony Barnosky's earlier prediction that we could be entering a mass extinction event in the next three centuries. Jackson revealed that he disagreed with Barnosky's assessment, and just as I was about to breathe easier, he hit me with this: Jackson thinks the intersection of climate change, invasive species, and ocean acidification is a recipe for serious disaster that is more urgent than Barnosky's predictions. "All these things could come together in a fifty- to one-hundred-year timescale to completely transform the surface of the earth," he said. "And once we realize it's happening, the results may come so fast that we can't stop them."

An ominous portent, I thought. But could another species make it better?

15

BEYOND *HOMO SAPIENS*

THE FOUNDING DIRECTOR of the Harvard Business School's Life Sciences Project, Juan Enriquez, posed a question at a TEDxSummit in Doha, Qatar, in April 2012. TEDx is the international component of TED, a nonprofit organization based in New York devoted to spreading game-changing ideas through short, powerful talks, often delivered at conferences. Enriquez's question was one that he had been enthralling audiences with a lot. He spoke about the history of life, tracing it from the Big Bang, to the birth of the stars, to the perimeter of the galaxy, to parts played by the sun, earth, and man, a history that spanned 14 billion years, involved trillions of stars, and then he asked the audience one question: What was the purpose of all this? He moved on to the next PowerPoint slide to provide the answer: A photo of Pamela Anderson and then Michael Jackson—the point being that man is the almighty purpose, the be all and end all of life, after which, he claimed, evolution flatlines to the end.

His next questions was "Wouldn't that be slightly arrogant? There has been something like twenty-five human species; why couldn't there be another?"

Indeed, why couldn't there, particularly if we were entering a mass extinction? Many scientists believe that natural selection operates mainly on the frontiers of change.

Seventy-seven thousand years ago, a human sat in a limestone cave in Africa on a cliff overlooking the Indian Ocean, cooled by a sea breeze and warmed by a small fire. He picked up a sharp rock and made a crosshatch design on a piece of reddish brown stone that scientists claim is the oldest known example of an intricate design made by a human being. It demonstrates the ability of man to communicate symbolically, which scientists believe sets *Homo sapiens* apart from other hominids on earth at that time.

This symbolic communicator with his stone tools and weapons had a competitive advantage as he moved out of Africa on the "Great Migration" into territory occupied by other species of the *Homo* genus. *Homo sapiens* first traveled to Asia eighty thousand to sixty thousand years ago. By forty-five thousand years ago this new hominid had settled Australia, Papua New Guinea, and Indonesia.

Keep in mind that by this point there may have been four different species on the planet: *Homo sapiens*, *Homo floresiensis*, Neanderthals, and Denisovans, the last a potential new human species described from a finger bone fragment found in Denisova Cave in the Altai Mountains of Siberia. But *Homo sapiens* eventually won out—the actual last man standing.

Harpending and Cochran think there has been significant evolution in the past fifty thousand years between human populations separated by great distance and geographical barriers. "No Finn could be mistaken for a Zulu, no Zulu for a Finn. There have been substantial changes in the genetic makeup of humans since man spread out of Africa, and those changes have taken on significant characteristics in different populations," they wrote in *The 10,000 Year Explosion*.

Robert Fogel, a University of Chicago economist, while studying the effects of American slavery, discovered that over the past few centuries—particularly in the last fifty years—Americans in general have been growing taller, living longer, and getting thicker. In 1850, the average American male was five feet seven inches tall and weighed about 146 pounds. By 1980 he stood five feet ten inches and weighed 174 pounds. A team of economists extended the statistical search worldwide and found the trend was global.

It turns out that advances in medicine, better nutrition, better working conditions, cleaner water, and a general reduction in pollution have netted humans a biological advantage. It's most dramatic when you consider age. When *Homo sapiens* first emerged in Africa about two hundred thousand years ago, the average life expectancy was twenty years. By the year 1900 it had become forty-four years. Today it is closer to eighty years, almost doubling in only a hundred years. And these are heritable trends passed on from parents to children, generated by improvements in health and medicine.

So is there another species in the wings?

As we've shown, nature does better when there are multiple species of any animal. Our single species is a bit unnatural. Multiple species of man have historically been the norm, rather than the single species we have today. Nature prefers biodiversity. A single species is not a strong holding position to maintain for any animal. But how could another species evolve?

There are two dominant types of speciation: allopatric speciation and sympatric speciation, with two other variants, peripatric speciation and parapatric speciation, in between. Allopatric speciation occurs in geographic isolation; sympatric speciation, however, occurs in the presence of other species, such as a lake where fish might segregate in the water column—some to the top, some to the bottom—and over time become separate species. Species could also form by specializing on different types of food, as the finches of Peter Grant's study on the Galápagos. Though Grant has yet to record a verifiable speciation, he has keyed into the possibility of that happening by studying how different sizes of seeds select for different beak sizes among the finches and how that selection pressure might produce another species.

Robert C. Stebbins studied populations of a small salamander at UC Berkeley in the 1940s. He proposed that *Ensatina* originated in the state of Oregon and spread south among the mountains on both sides of California's Central Valley, the valley floor being too dry and hot for salamanders. As the pioneering populations moved southward, they evolved into several subspecies. Each subspecies had new color patterns as well as adaptations for living in different environments. But by the time they met again at the southern end of the Central Valley, they had evolved so much that they no longer interbred—even though they blended into one another and had interbred in the mountains around the valley rim. This minimum of isolation was all it took to differentiate a species.

Are we overstating geological separation as a starting point for the next species?

Could we do it with sympatric speciation—splitting into two in the presence of others? What evolutionary pressures in the human population might produce another species of modern man? Cochran and Harpending have looked at how cultural isolation altered the genetic code of some of our ancestors. The two scientists speculate that European Jews were genetically isolated during the Middle Ages, not by oceans or mountain ranges, but by Jewish rules against intermarriage reinforced by external prejudices against Judaism. For most of the Middle Ages, intermarriage with non-Jews, as well as conversions to Judaism, were quite rare.

The dominant trait that puts them at a culture difference—as being large does the Samoans, as being tall does the Tutsis, and as being milk tolerant does the Scandinavians—is intelligence. Several different sphingolipid mutations—a special type of genetic anomaly that creates a buildup of lipid or fat molecules that enhance signal transitions in neural tissues—could cause this. This increases connections in the neurons, the basis of the central nervous system.

Ashkenazi or Eastern European Jews in the Middle Ages specialized as financiers, estate managers, and merchants, jobs that required analytical thinking along with cultural understanding, since they often served as intermediaries between Christians and Muslims. These effects were highly heritable, each generation becoming slightly more adapted and slightly more analytical than the first.

As a result, European Jews have one of the highest IQs of any ethnic groups known, claim Harpending and Cochran. Their IQs average around 112 to 115, whereas other Europeans average about 100. But because of a lack of diversity in their genes, they also have a set of genetic diseases a hundred times more often than other Europeans. These include Tay-Sachs disease, Gaucher disease, familial dysautonomia, and a couple of forms of hereditary breast cancer.

As with resistance to malaria, there are pros and cons. European Jews are subject to some serious bodily ailments, yet their brains appear to work better. Their numbers of prominent scientists are ten times greater than their percentage of the populations of the US and Europe. In the past two generations, they have won more than a quarter of all Nobel Prizes for science, although they make up less than one six-hundredth of the world's population. It appears that cultural isolation while dealing with difficult white-collar occupations—long-distance trade, managing ranches and estates, collecting taxes—among volatile cultures is great training for math and science careers.

Could sympatric evolution be achieved by another formula? Harvard's Enriquez says that isolation could be achieved by what he calls the "sexy geek syndrome." This might occur if computer programmers are put in isolation and interbred. Such a scenario already exists at Google headquarters in Mountain View, California. Nicknamed Googleplex, it's like a college campus filled with employees wearing jeans, wandering around among the dogs, bicycles, and volleyball courts. Google sends out special luxury buses to pick up its workers, which prevents pairings with non-Google people on the drive to work.

Googleplex buildings have high ceilings, lots of natural light, and open cubicles for offices. There are a number of cafeterias where employees can sit around tables, argue algorithms, listen to rock music, and eat free gourmet food. You can even bring your dog to work. Google guards its employees jealously, offering them high compensation, a work environment that provides all their needs, and a system that allows employees to work about one day each week on their own projects, giving everyone the opportunity to be the next Larry Page or Sergey Brin, the founders of Google. How do you walk away from that?

Is this enough isolation to achieve speciation one day? Perhaps—especially since many computer jobs require twelve-hour days, which in itself limits the number of hours one can search for a mate.

We've proven we can change the genetic makeup of plants and animals, so how about us? We don't need to wait for natural selection: we can start selecting right now. The cost of genomic sequencing, the key to moving modern medicine from reactive standards to personalized prevention, has fallen astronomically. When the Human Genome Project was announced in 1990, deciphering the genome of one man was budgeted at $3 billion. By 2001 the cost was down to $3 million. In 2010 it was below $5,000. By 2012 it was below $1,000. At this rate, in ten years a fully sequenced human genome should cost about $10.

As genetic screenings become more common, designing the body to alter genetic weaknesses will be more common as well. Angelina Jolie getting a double mastectomy because of a gene in her body that makes her more susceptible to breast cancer is just the start. It may one day be possible to change the gene rather than the result. The negative aspect is that many genes perform more than one function. Changing a gene to match a given result may have unintended consequences. Trial and error will be necessary here.

What will be the big forces behind genetic manipulation? The University of Washington's Peter Ward sees parents as strong selective forces, since many will want their offspring to live long, look good, and be brainy. "If the kids are as smart as they are long-lived—an IQ of 150 and a life span of 150 years—they could have more children and accumulate more wealth than the rest of us," wrote Ward in a January 2009 article for *Scientific American*. Socially they would be drawn to others of their kind, which could lead to speciation.

Parental desires could provide the big necessary push for the creation of designer genes if only to ensure that their children will be talented, the right height, or the right weight. Such considerations could be a major force for not just designer genes but designer children. Stanford University's Rob Jackson speculates, "What would happen if women could order Brad Pitt's sperm from the back of a magazine? Even better, what if they could mix Will Smith's smile and George Clooney's eyes from a catalog? It will fundamentally change the human race."

What if we could alter male genes to make the perfect soldier? According to Henry Harpending, "The Chinese talk about that often—without batting an eye." The perfect soldier . . . what about the perfect nuclear physicist?

Each of our cells contains our entire genome. Every one of your cells has the genetic blueprint to make an entire you. In 2009, Chinese scientists took skin cells from a mouse and turned them into stem cells. Then they took those stem cells and allowed them to regrow, differentiate, and give birth to a live mouse. Which could then reproduce normally.

The mouse, Xiao Xiao ("Tiny" in Mandarin), was born from one of its mother's skin cells. What this means is that, in theory, it should be possible to take any one of our cells and create a clone. Remember Dolly, the cloned sheep? Though society has frowned on repetitions of cloning, how long will it take before someone decides they are so special that there needs to be more than one of them? The ability to clone ourselves from skin cells, to change our organs at will, could lead to an explosion of hominid species.

UPLOADING THE MIND

There are other variations on this copying thing. One is uploading your brain. Ed Boyden, a synthetic neurobiologist at MIT, is currently attempting to map the human brain. There are more than 100 billion computational elements in our brain. So Ed designed his own way of isolating brain circuits. He learned how to take stuff out of algae and use it to illuminate and activate specific brain pathways. He can then use the light to watch what is happening inside a brain as a mouse moves an arm, sees, touches, or smells something.

To get beyond simple neural mapping, I hopped a fast train from London Paddington to Oxford station and then took a cab that circled the Oxford campus through stately old buildings that I recognized from watching British mysteries, finally arriving at the Future of Humanity Institute at the edge of the Oxford University campus.

Nick Bostrom, a confident, cerebral man of medium height and slim build, met me in his second-story office overlooking the historic city. Bostrom likes to spend his time contemplating the various threats to our existence—their probabilities and what we might do about them. Bostrom thinks there is a big gap between the speed of technological advance and an understanding of the dangers it has for man.

That overcast day, however, we were talking about the possibilities involved in uploading one's mind, though Bostrom didn't ignore its dangers. Bostrom believes that as technology accelerates, "at some point the technology of mind uploading becomes available and we may transform the human brain into software." He said this could be possible by making high-resolution scans of thinly sliced layers of the brain and then uploading those scans to a computer. He felt it was not that far away.

The idea is that human consciousness could well exist in a machine after the body has been discarded or, more likely, outlived. "The main pressure for this could come from people who are terminally ill and want to try immortality," Bostrom said. He thought our neural architecture might exist on a computer, but our consciousness might "reside in a robot in the real world or as an avatar in a virtual reality."

There is precedent for this in the computer game world. Second Life is a 3-D online community that has millions of users who take regular walks in a virtual world where everyone is beautiful. It is the creation of Linden Lab in San Francisco and provides its users with a real-time experience on their personal computers, allowing them to wander around castles, deserted islands, other fantastic 3-D environments, and meet thousands of online participants, talk, and even have simulated sex. The company reports that the average player spends about twenty hours a week in these environments.

Bostrom claimed that once society is uploaded, it would be practical to separate our abilities into nodules that could perform different tasks. After all, it would be more efficient to hire a math nodule rather than waste a lot of time doing math. One of the goals of artificial intelligence—to make all knowledge accessible to all people—could be accomplished much more easily if we were all connected pieces of software. And Bostrom feels this will naturally breed specialization.

Once specializations are standardized, copying oneself would become logical, because it would increase the worth and the assets of the individual. He said that there would still be people who would like to do things themselves—for example, hobbyists who would enjoy planting vegetables or knitting their own sweaters—but they would be outcompeted by people that didn't need such things. The old argument that man needs rest and relaxation once in a while would disappear in an uploaded world, since software packages don't need to rest.

In such a world, Bostrom sees simulated life splitting into two groups. One group would replicate current human values by engaging in such enjoyable stuff as humor, love, game-playing, art, sex, dancing, social conversation, food, and the like. Though Bostrom thinks those activities may have been adaptive in our evolutionary past, he wonders if they would be adaptive in the future. "Perhaps what will maximize fitness in the future will be nonstop high-intensity drudgery—work of a drab and repetitive nature—aimed at improving the eighth decimal of some economic output measure," said Bostrom.

The competing superpowers of Bostrom's future uploaded world would be the all-work-and-no-fun group, or what he calls the "fitness-maximizing competitors," versus the "happiness and well-being group." He envisions the fitness-maximizing competitors as eventually taking over the capital of the day from the happiness-and-well-being group, since the latter might still like to play now and then—an activity okay for today's organic brains, but unnecessary, time wasting, and fruitless to our future software selves.

This would result in a future world where everyone is a fitness-maximizing competitor or where some happiness-and-well-being agents continue to survive, but their activities would go on underground.

Bostrom thinks that if we want to continue with this interest in occasional happiness, we might need to pass laws that tax fitness-maximizing activities while subsidizing happiness-and-well-being cognitive architectures. For example, some fraction of our resources might be set aside in a happiness-and-well-being conservation fund. We might also have to pass laws against building artificial intelligences that are hostile to human values—another of Bostrom's worries.

THE AI PROBLEM

James Barrat, author of *Our Final Invention: Artificial Intelligence and the End of the Human Era*, also worries about runaway AI. Part of this results from the fact that national defense institutions are among the most active investors in AI. Their emphasis gives a competitive edge to whoever can advance the furthest and the fastest, and much of their research is designed to kill humans.

Computers have grown exponentially. Still, there are enormous hurdles to overcome for advanced AI on a level with human intelligence to become a reality. Visual recognition software currently can't tell the difference between a dog and a cat. Using AI to diagnose ailments could be a tremendous advantage, since the computer can access so much more data and so much faster than a physician, but if vision is part of the diagnosis, a computer might analyze all symptoms and not notice there was a bullet hole in the patient.

But those things will eventually work out, whether through advanced computing or by reverse engineering of the human brain, copying its biologically inspired methods rather than trying to duplicate its results with a different technology. The worry is that once advanced machines are in place, their own progress will be turned over to the other AI computers and they will grow exponentially.

The turnover of power might seem gradual, painless, and fun, but the consequences could be fatal. They might treat us as the boss or they might treat us as we treat our primate ancestors: monkeys, apes, and orangutans, who've been kept in zoos, used as laboratory animals, and whose wild populations are all endangered, with little chance of surviving into the distant future. With advanced AI we will have introduced another species to our planet that could outcompete us.

Bostrom's simulation argument takes into consideration the idea that we may already be in a virtual reality. He says it's one of only a few possible descriptions for real life today. Civilizations like ours will go extinct before they can create virtual worlds run by uploaded minds. Or they will lose interest in creating computer simulations detailed enough that the simulated minds within them would be conscious. Or, lastly, we're almost certainly living in a computer simulation. You're an avatar and so am I.

BACKING AWAY FROM THE CLIFF

But what if society doesn't want to upload? What if slicing your brain up like a baked ham hits the market and tanks? What if the computers get hung up on the dog-versus-cat problem? Perhaps we should just continue our ways, procreating, consuming resources, and hope for the best.

Georgii Gause pondered the options of "continuing our ways" some time ago. While a student at the University of Moscow in the 1920s, he performed a classic experiment in which he placed half a gram of oatmeal in water, boiled it to create a broth, and poured the concoction into small, flat-bottomed test tubes. Into each of those tubes he placed a small amount of two different single-celled microorganisms, one species per tube. Each test tube was a unique ecosystem, a food web with a single mass of tissue. He then set them aside in a warm place for a week and came back to review the results. His book, *The Struggle for Existence*, published in 1934, detailed these findings.

At first the number of organisms grew slowly. On a graph with time on the bottom and numbers of microorganisms on the side, the line rose gradually as the number of organisms increased. But the line hit an inflection point, and the number of organisms exploded as the line rose suddenly steeper. The frenzied expansion of organisms continued skyward until they exhausted their food, after which the line leveled and then plummeted as the organisms began to die, the line plunging toward zero.

If we look at the growth of *Homo sapiens*, there are similarities between the ongoing human population experiment and that of the microorganisms that Gause described in the 1930s. If geneticists are correct, *Homo sapiens* exited Africa with no more than a few hundred people and migrated over the world, enjoying a series of Edens. Yet, at our greatest expansion, we barely numbered five million. But then about ten thousand years ago we invented agriculture, and the human population began to rise steeply.

In the nineteenth and twentieth centuries we discovered ways to thicken grain stalks, along with methods of producing fertilizers and better systems of irrigation, and in just two hundred years our population went from one billion to seven billion, with two or three billion more on the way. The bulk of this growth has been in the last fifty or sixty years, a single lifetime. We are definitely like a virus—or at least a single-celled organism. Part of the reason for the frenzied rise in Gause's experiment was the lack of competing species in the test tube. Competing species would have slowed things down, formed buffers, created competition. But we're eliminating that possibility . . . aren't we?

In 2000, Dutch chemist Paul J. Crutzen gave a name to our time: the "Anthropocene," the age of man. He regards the influence of human behavior on the earth's atmosphere as so significant as to constitute a new geological epoch.

But is there a limit to this influence? What if, like Gause's microorganisms, our numbers rise to an apex but then plummet back down? What if overpopulation, starvation, disease, and the obliteration of native species aren't healthy and we end up with perhaps 20 percent of the numbers of our current species—back to maybe a population of one or two billion, where we were just two hundred years ago—although isolated perhaps by the aftermath of the devastation?

Ian Tattersall, curator emeritus at the American Museum of Natural History in New York and author of *Masters of the Planet: The Search for Our Human Origins*, doesn't believe in the emergence of another species under present conditions. "Man has spread all over the world, and mass transportation has made it too easy for us to intermingle," says Tattersall. But when we propose the idea of a crash similar to what Georgii Gause witnessed in the petri dishes, Tattersall says, "Then all bets are off."

If man's future development mirrors Gause's experiment, then it might be possible for a member of *Homo sapiens* to develop into another species in a world with far fewer of us present.

The next species could arise out of isolation created by mass extinction as well. It could also be isolated within the present modern cultural context, because of cultural stigma and prejudice. Or genetic engineering could create a "superspecies" that finds it repugnant to mate with "less-than-super" beings.

Coexistence with the next species must begin by recognition: knowing who they are and what they look like. And how would we do that? The difference could start with something as simple as a few genes that could help the next species consume some more efficient food, as the Indo-Europeans did, or make men more resistant to the future plague of diseases, as were the conquistadors. However, neither the Indo-Europeans nor the conquistadors were different species, so we'd need more.

Science warns against the utopian view that selection will continue to upgrade nature much like a new cell phone. Selection promises only that a new species will be able to outcompete its ancestors in a given region at a given time. Evolution might create a wiser individual with a better, more long-term perspective of the world, but it would happen only as a side benefit; the primary benefit would address maximum fitness right now.

Still, suppose this could happen. Is this around the corner, far off in the distance, or has it already happened and we just don't know it? If you were to take a Neanderthal, clean him up, give him a haircut, put him in some new clothes, and shove him out onto the street, few would recognize him as out of place. Douglas Palmer, in *Origins: Human Evolution Revealed*, claims that Neanderthals were a little broader than modern man, the brow a little more pronounced, the arms a little stronger, but he could be a track star, a football player, or a character actor for all we know.

The next species might be a little leaner, continuing the trend from past hominid development, with a larger head for a larger brain, and with a diminished nose, brow ridge, and other facial features. Still, could you pick the next species out of a crowd any better than a Neanderthal? And—get this—even if *you* are the next species and you're reading this, trying to get a better understanding of those you are about to replace: you're doomed, too. As Smithsonian's Hans-Dieter Sues says, 99.99 percent of all the species on earth have gone extinct. There is no reason to think that man—or his immediate descendants—will, in the long run, outlive them all.

Time is part of the problem. Man has difficulty contemplating the enormous range of times that have passed in the history of our planet and how our own history compares with that. "Deep time" is what the scientists call it, and man is, apparently, not a "deep" thinker. Paleontologist Stephen Jay Gould impressed this on his Harvard students by giving them this simple analogy: if our planet's beginning is the end of your nose and its present your outstretched fingertip, then a single swipe of a file on the finger's nail wipes out all human history. This doesn't refer to the brief period of human civilization but the entire existence of *Homo sapiens* and all his hominid ancestors.

Scientists also like the clock metaphor, which gives human time another perspective. If we look at the 4.5-billion-year history of our planet in terms of a twenty-four-hour day, then the Cambrian explosion doesn't occur until about 10 p.m. Dinosaurs don't appear until after 11 p.m. and the big asteroid marks their end at twenty minutes before midnight. Man doesn't appear until the last few seconds.

Man just might be the ultimate live-fast-die-young animal. Consider that the average longevity for a mammalian species is only one to two million years. We've been around only one-tenth of that, about two hundred thousand years, and our existence is gravely threatened. Sam Bowring at MIT thinks the paleontological record for man will expose but a thin layer of metals that *Homo sapiens* dug out of the ground.

Jan Zalasiewicz, lecturer at the University of Leicester, UK, and author of *The Earth After Us*, admits that it is hard to compare human and geological timescales. He suggests we make a trip to the Grand Canyon near Flagstaff, Arizona, and look down into the mile-deep chasm, whose strata span 1.5 billion years. "Measured on this scale, our own species would fit into a layer about three inches thick, while our industrial record would be confined to about one-hundredth of an inch," he says. Such an interval would seem almost instantaneous to a geologist—not much more than a meteor strike.

But the human population boom, though bad for nature, will be great for future interplanetary geologists. The more specimens, the more chance for fossils, and man has come up with an explosion of specimens in just the last hundred years. Perhaps our best choice is to aim for fossil importance.

But if you really want your bones to end up in a museum diorama somewhere else in the galaxy, then, as they say in real estate, it's all about location, location, location. You need to take your final bow down by the seashore or at a river mouth where sediments, soil, and rocks are actively being deposited upon one another, earth upon earth, in layers. For if you sing your final swan song on top of a cliff or a mountain, no matter its dramatic appeal, that's not good for fossil preservation: there's too much active erosion. Gradual accumulation or "deposition" of soils, as paleontologists refer to it, layer upon layer on top of your bones, is far better for the preservation of future new fossils than erosion.

So how might *Homo sapiens* finish their act, and nature move to center stage? Look at the way Neanderthals did it. It was in the caves at the base of the Rock of Gibraltar, at the tip of the Iberian Peninsula, that the last evidence of the Neanderthals was found. Global temperatures were cooler and a lot of water was locked up in ice, lowering sea levels 80 to 120 meters. This opened up a huge portion of the coastal shelf that today is under the Mediterranean Sea.

There was plentiful meat and game but not enough water. The end could have come during a summer drought when life was stretched to its limit. In that environment, at that time, rains did not come in summer, and in some years they did not come at all.

Extinction comes when a species reaches a point when births do not keep up with deaths, and numbers gradually diminish. Such a condition for man may come by the end of the century, when many demographers predict population growth may start to diminish. This could be a great thing: man finally reeling in population growth. But is it cause for celebration, or the beginning of the end? If future populations don't rein in resource use, then diminishing population growth may be an empty promise.

In recent years, there have been a number of convocations at major universities and governmental offices on threats to man's existence. Asteroids and comets always come up, since an asteroid had lots to do with the end of the dinosaurs. An asteroid or comet could do us in, as we learned when scientists turned their telescopes to watch the impact of the comet Shoemaker-Levy 9 on Jupiter in July 1994. The warning for such events may be only a year, and the consequences—had that comet hit Earth—could certainly have taken out our species; but it's more likely to take only a portion of us, with the rest recovering, as in the years following World War II, in a relatively short time.

Sudden climate change could do it, but we've already survived the Younger Dryas event, as well as the last two glacial epochs, and we're still kicking. Climate change is more apt to wreak havoc on nature, and nature to wreak havoc on us secondhand. Biological warfare? Runaway nanotechnology? We made it through the Black Plague. Even Native Americans survived conquest by the Europeans.

Robots armed with some sort of super–artificial intelligence, and smarter than their makers—able to access the world's knowledge from the cloud—might be amazingly helpful or harmful, depending upon their designers, but are not likely to turn on all of us at once.

Thermonuclear power has lots of potential. In 2012 the *Bulletin of the Atomic Scientists* announced that it had moved its Doomsday Clock forward to five minutes to midnight. In 2010 the clock had been pushed back to six minutes to midnight, but it was moved closer again based upon the fact that arms reductions worldwide have stalled, as have efforts to curb climate change. The *Bulletin* mentioned that we still have over nineteen thousand nuclear weapons, "enough to destroy the world's inhabitants several times over," and that many countries are in the process of upgrading their current arsenals. The complexity and resources involved in making these weapons has slowed their spread, but what happens if someone should discover a way to make them from sand—or something similar?

According to the United Nations Population Division, the median population scenario, often seen as most likely, predicts that by 2050 the world will have 9.2 billion people. That's up from earlier estimates due to increased fertility in Europe and the United States. In 1800, only two hundred years ago, there were only one billion of us.

Man is immensely resourceful at extracting our natural resources from the earth. As Charles Mann, author of *1491*, says, "It is our greatest natural blessing. Or *was* our greatest natural blessing." We are getting close to the end in some vital areas. In the next hundred years, despite improving extraction technologies, we could run out of oil, phosphorus, and perhaps even fertile land. The World Bank predicted in the late twentieth century that most twenty-first-century wars would be fought over water supplies.

According to UC Santa Cruz professor Jim Estes, "We will either evolve into something new, or we will become a dead end. This *Homo sapiens* lineage will cease to exist. The key thing is the next fifty to one hundred years; that's the big question. And what's the quality of life going to be like for those that live through that period? Beyond that, our ability to forecast is very poor. There could still be humans, as we know them today, fifty thousand years into the future. But certainly a million years into the future . . . we will be gone."

But what might hold us back from that inflection point that Georgii Gause described in *The Struggle for Existence*, where the line plummeted down the other side of the steep rise? The answer is: Don't go there. Hold back. Stop! But in making that request, we are asking humans to do something that no other species has ever done: constrain its numbers voluntarily. It's a gargantuan order. Zebra mussels in the Great Lakes, brown tree snakes in Guam, water hyacinth in African rivers, Burmese pythons in Florida—all continue to try to overrun their environments.

But Charles Mann expresses hope. In a recent article for *Orion* magazine, he described the conditions of slavery in the eighteenth and nineteenth centuries and how society evolved away from the practice. This was also about the time that *Robinson Crusoe*, Daniel Defoe's famous novel, came out depicting Crusoe and his men shipwrecked on an uninhabited island off Venezuela where they learned to live off the land. Defoe made Crusoe an officer on a slave ship, then an honorable occupation. When the book came out in 1719, no one complained. Slavery was accepted then.

But in a few decades in the nineteenth century, slavery almost vanished. The road to this change in terms of human consciousness is enormous. In 1860, slaves, all told, were the single most valuable assets in the United States, worth about $10 trillion in today's money. But the tide turned on slavery, though at great cost to individual lives and national finances. The American Civil War killed more than six hundred thousand combatants and wrecked the US economy, but slavery died. And it didn't just die in the US: in the nineteenth and early twentieth centuries it died in Great Britain, the Netherlands, France, Spain, Portugal, Korea, Russia, China, and quite rapidly in most of the rest of the world. There are still some vestiges of slavery in the world—forced labor, sexual slavery, and indentured servitude—but there are few open markets for slavery, and for the most part, nations don't thrive on their slaves.

Another example of a great change of attitude is the rise of women. As the Civil War raged in the US to free the slaves, women were also denied essential rights. In both the North and the South, few women could attend college, hold public office, run a business, or vote. Men dominated women in every society. Voting rights for women were not extended until the first half of the twentieth century in most nations. In the United States, women's suffrage was achieved gradually, reaching nationwide status in 1920 with the passage of the Nineteenth Amendment to the US Constitution. Today women in the US comprise the majority of the workers and the majority of the voters. And this is the same in many countries all over the world.

Something similar is now happening with gay, bisexual, and transgendered people. Major legislative and legal changes are being made, and an attitude of acceptance is growing.

Enormous tidal waves of change are still possible for the human species.

Stopping man from killing himself will take more than behavioral modification. Like a world of dieters fending off hunger, we would have to push back from the table of reproduction, renounce growth, and limit our use of natural resources in order not to hit that fatal inflection curve—to avoid the catastrophe of nature making those selections for us.

According to Peter Ward, the University of Washington professor and paleontologist, earth is presently in what he calls the habitable zone, the ideal distance from the sun. Astronomers looking for habitable planets in the galaxy look for those that are similarly distant from their stars. If a planet is too close to the sun, it gets too hot. The surface of Venus is hot enough to cook dinner. If it ever had water, it has evaporated into space. On Mars, all the water is frozen. Both Mars and Venus are outside the habitable zone. The trouble is the habitable zone moves outward with time. This is because our sun grows hotter with age. Under these circumstances, the earth will move outside of the habitable zone in 500 million to one billion years. Life may exist on earth, but it will be microscopic. Mars might be a good bet then. If one can wait. If we last that long.

Ward thinks that man will survive the distance, "but the animals and plants along for the ride on this planet that we cockily co-opted will not be so fortunate." The future of the planet may be permeated with episodes in which mankind is every once in a while knocked back to the Stone Age.

Bill Schlesinger, president of the Cary Institute of Ecosystem

Studies, says, "The conditions of our planet are largely determined by the biosphere—the collective action of all the species on earth. These species control the composition of our atmosphere and oceans, the climate, and the total amount of plant production on land and in the sea, upon which we all depend for food, fuel, and fiber. It's hard to believe we could make it without them."

No single cause will take humans out. But multiple causes have a chance. In the end it may be as Douglas Erwin describes the end of the Permian—his so-called *Murder on the Orient Express* theory—in which a multiplicity of causes created the Permian extinction. Such a multiplicity of causes may have provoked the Cretaceous extinction as well: large reptile herbivores mired in a long-term decline, plus the effects of the Deccan Traps, one of the largest volcanic provinces in the world, about half the size of India. Vincent Courtillot from the Institut de Physique du Globe de Paris says, "It released ten times more climate-altering gases into the atmosphere than the nearly concurrent Chicxulub meteor impact."

The sixth mass extinction may also be multicausal, arriving on various tracks, including overpopulation, runaway climate change, unbridled disease, and a planet that runs out of modern man's necessities.

Might we live on with nature as robots after uploading our minds? Perhaps. But that's not as sure a bet as one placed on nature. Nature will survive. Life will continue, though in different forms, different species. Ecosystems will recover and thrive one day as they did before us, with different sets of players, perhaps different sets of rules.

With the heavy cloak of humankind laid to rest, nature may take one enormous sigh of relief, and then press on, to recover its former glory.

Acknowledgments

I HAD A LOT of patient, eloquent, and informed help with this book for which I am truly grateful:

Tom Schulenberg at the Cornell Lab of Ornithology and Louise Emmons at the Smithsonian let me accompany them and their magnificent crew—Lawrence López, Mónica Romo, Brad Boyle, and Lily Rodríguez—on my first visit to the cloud forests of the eastern Andes. Miles Silman at Wake Forest University let me visit that place again.

Anthony Barnosky at the University of California, Berkeley, first warned me of the possibility of our own mass extinction, though I would hear others such as the famed anthropologist Richard Leakey say the same.

Guadalupe Mountains National Park geologist Jonena Hearst led me up a West Texas trail to the Capitan Reef and showed me the fossils from the Permian mass extinction. Harvard's Andy Knoll and MIT's Sam Bowring gave it an atmospheric and geological perspective.

William Schlesinger at the Cary Institute of Ecosystem Studies told me about early life on earth and what part chemistry played in its appearance. Rick Ostfeld, also at Cary, told me how the loss of forest and animal species was a setup for disease. Rob Jackson at Stanford took me down into Powell's Cave in Central Texas to show how juniper, an invasive plant, was stealing water from the Edwards Plateau.

Stuart Findlay at Cary helped me understand New York's Catskill/Delaware Watershed. Dalia Amor Conde of the Max-Planck Odense Center showed me how jaguars and rain forest are vital to human life in Central America. Stan Smith at the University of Nevada, Las Vegas, and Greg Okin at the University of California, Los Angeles, showed me how desert crusts are vital to the longevity of the American desert. The paleontologist Charles Marshall at the University of California, Berkeley, gave me an Australian perspective of US species loss.

Kevin Coleman and Paul Poulton at Rothamsted Research in the UK described how important our soil is to the future of agriculture. Dan Richter at Duke University took me to my first conference on soil. Eduardo Neves at the Universidade de São Paulo introduced me to *terra preta*, black earth, an earthy gift from ancient Amazonian Indians.

Leslea Hlusko at the University of California, Berkeley, and Jackson Njau at Indiana University let me share their field site at Olduvai Gorge in Tanzania. Miriam Ollemoita took me to her Maasai village. And Tomos Proffitt entertained me by making Stone Age hand axes.

Hans-Dieter Sues at the Smithsonian explained how crocodiles were the real dominant predator during the Triassic. Rick Potts, also at the Smithsonian, talked of the role climate may have had in the development of early man. Kathy Weathers at Cary turned me on to the fog forests of eastern Chile.

Thanks to Jim Estes at UC Santa Cruz for helping me understand how killer whales could turn to sea otters for food; to William Gilly at Stanford for all the time he spent teaching me about Humboldt squid; to Gretchen Hofmann at the University of California, Santa Barbara, for her insight on ocean acidification; and to Frank Hurd for his tireless efforts to bring clean, sustainable aquaculture to the natives of Baja California to make up for the disappearing fish.

Felisa Smith at the University of New Mexico showed me how animals once got very big, and how they might get there again. Blaire Van Valkenburgh at UCLA showed me how vital and diverse nature once was four miles south of Hollywood.

Henry Harpending at the University of Utah convinced me that man is still evolving. And Oxford's Nick Bostrom delighted me with his predictions of the future of man on earth.

Stephen W. Smith at Duke University, author of sixteen books, was my mentor throughout, encouraging me to write faster and encouraging me to go to Africa. Thanks to Jim, Julie, Noel, Eddie, Geary, and the other members of the Sky Valley hiking club with whom I spent many mornings and whose questions kept me thinking. Thanks to my other writer friends, Susan Squire, Stewart Weiner, and Susan Vreland, for all their sage advice.

I owe a lot to my family, Bill, Barbara, Sam, Ann, Marguerite, and my wife, Maggie, for putting up with me during this long process. Also to Mike LaJoie, Alex Wexler, Grace Murphy, and Gary Kott for their invaluable support.

Thanks to Jud Laghi for the fire and enthusiasm he displayed in representing me. Also to Hilary Redmon, who first saw the promise in it. Thanks also to Leah Miller, Webster Younce, and Karen Marcus for herding me through the early days, and Sydney Tanigawa, editor extraordinaire, who helped me wrangle a sometimes disjointed manuscript into its present form.

Thank you all.

Notes

Prologue WE HAVE NO IDEA WHAT WE'RE IN FOR

1 *It was mid-morning, June, during the tropical dry season*: Michael Tennesen, "Expedition to the clouds," *International Wildlife* (March/April 1998), 22–29.

1 *the group of celebrated biologists*: L. E. Alonso, A. Alonso, T. S. Schulenberg, and F. Dallmeier, eds., *Biological and Social Assessments of the Cordillera de Vilcabamba, Peru* (Washington, DC: Conservation International, Center for Applied Biodiversity Sciences, 2001).

2 *the most biologically diverse*: Norman Myers, et al., "Biodiversity Hotspots for Conservation Priorities," *Nature* 403 (February 24, 2000), 853–58.

2 *during past ice ages*: Mark B. Bush, Miles R. Silman, and Dunia H. Urrego, "48,000 Years of Climate and Forest Change in a Biodiversity Hot Spot," *Science* 303, no. 5659 (February 6, 2004), 827–29.

3 *"shoestring distributions"*: Michael Tennesen, "Uphill Battle," *Smithsonian* 37, no. 5 (August 2006), 78–83.

4 *a sixth of the world's plant life*: Author interview with Tom Schulenberg, July 2013.

5 *The palpable haste of modern biologists*: Anthony D. Barnosky, et al., "Has the Earth's Sixth Mass Extinction Already Arrived?" *Nature* 471 (March 2, 2011), 51–57; Norman Myers and Andrew Knoll, "The biotic crisis and the future of evolution," *Proceedings of the National Academy of Sciences* 98, no. 10 (May 8, 2001), 5389–92.

5 *It took* Homo sapiens *less than 200,000 years*: www.pbs.org/wgbh/nova/world balance/numb-nf.html.

5 *history of earth as a twenty-four-hour*: http://www.geology.wisc.edu/homepages /g100s2/public_html/history_of_life.htm.

6 *Recoveries followed all the mass extinctions*: Douglas Erwin, *Extinction: How Life on Earth Nearly Ended 250 Million Years Ago* (Princeton, NJ: Princeton University Press, 2006), 1–30; Richard Leakey and Roger Lewin, *The Sixth Extinction: Patterns of Life and the Future of Humankind* (New York: Anchor Books, 1996), 39–58.

6 *"From the wreckage of mass extinctions"*: Erwin, *Extinction*, 15.

7 *"With them, Earth's biodiversity remains"*: Barnosky, et al., "Has the Earth's Sixth Mass Extinction Already Arrived?" *Nature* 471 (March 3, 2011), 51–57.

8 *"Virtually 99.999 percent of all life"*: Author interview with Hans-Dieter Sues, April 16, 2012.

1 A MASS EXTINCTION: THE CRIME SCENE

11 *The Capitan Reef, though long dead*: National Park Service, Geology Field Notes, Guadalupe National Park, Texas, http://www.nature.nps.gov/geology/parks/gumo.

13 *enormous depression known as the Delaware Basin*: Author interview with Jonena Hearst, November 6, 2012.

13 *During the Permian period this gallery of life*: Erwin, *Extinction*, 2–7.

15 *It hasn't been that long since man*: "Fossils and the Birth of Paleontology: Nicholas Steno," Understanding Evolution, University of California, http://evolution.berkeley.edu/evolibrary/article/history_04.

15 *The awareness of fossils grew*: "William Smith (1769–1839)," University of California Museum of Paleontology, www.ucmp.berkeley.edu/history/smith.html.

15 *Geologists discovered that layers of rock in North America*: Jonathan Weiner, *The Beak of the Finch: A Story of Evolution in Our Time* (New York: Vintage Books, 1995), 109.

15 *These upheavals presented*: Stephen Jay Gould, *The Structure of Evolutionary Theory* (Cambridge, MA: Belknap Press of Harvard University Press, 2002), 745.

16 *Perhaps the most famous of the five extinction events*: Leakey and Lewin, *The Sixth Extinction*, 47–56; http://www.nobelprize.org/nobel_prizes/physics/laureates/1968/alvarez-bio.html.

17 *a million times more energy*: "Experts Reaffirm Asteroid Impact Caused Mass Extinction," University of Texas at Austin, http://www.utexas.edu/news/2010/03/04/mass_extinction/ March 4, 2010.

17 *huge tidal waves spread*: Peter Ward, *Future Evolution: An Illuminated History of Life to Come* (New York: Times Books, 2001), 24–26.

17 *The first vertebrates*: Author interview with Jonena Hearst; "Geological Time: The Permian; Terrestrial Animal Life and Evolution of Herbivores," Smithsonian National Museum of Natural History, http://paleobiology.si.edu/geotime/main/htmlversion/permian2.html.

18 *best studied Permian-Triassic boundary sequences in the world*: Erwin, *Extinction*, 83.

19 *This eruption occurred about 252 million years ago*: Seth D. Burgess, Samuel Bowring, and Shu-zhong Shen, "High-precision timeline for earth's most severe extinction," *Proceedings of the National Academy of Sciences* 111, no. 9 (February 10, 2014), 3316–21.

19 *The end result of the buildup of* CO_2: Scott Lidgard, Peter J. Wagner, and Mathew A. Kosnik, "The Search for Evidence of Mass Extinction," *Natural History* (September 2009), 26–32.

20 *Floods skipped across the earth*: Erwin, *Extinction*, 144–45.

21 *In a 2007 paper in* Earth and Planetary Science Letters: Andrew H. Knoll, et al.,

"Paleophysiology and end-Permian mass extinction," *Earth and Planetary Science Letters* (2007), 295–313, www.sciencedirect.com.

21 *30 percent of the species of plants and animals*: Author interview with Andrew Knoll, April 24, 2012.

23 *the arrival of man*: Guy Gugliotta, "The Great Human Migration: Why humans left their African homeland 80,000 years ago to colonize the world," *Smithsonian* (July 2008).

2 ORIGINAL SYNERGY

24 *chemistry is often underrated*: William Schlesinger and Emily Bernhardt, *Biogeochemistry: An Analysis of Global Change* (Waltham, MA: Academic Press, 2013), 20–32.

25 *"the road to life on planet Earth"*: Author interview with William Schlesinger, October 16, 2011.

26 *Alexander Oparin independently suggested that all the ingredients for life*: http://biology.clc.uc.edu/courses/bio106/origins.htm.

26 *In the 1950s, Stanley Miller*: Nick Lane, *Life Ascending: The Ten Great Inventions of Evolution* (New York: W. W. Norton, 2009), 8–33.

27 *Possible solutions emerged*: Toshitaka Gamo, et al., "Discovery of a new hydrothermal venting site in the southernmost Mariana Arc," *Geochemical Journal* 38 (2004), 527–34.

28 *Around the turn of the twenty-first century*: William Martin and Michael J. Russell, "On the origin of biochemistry at an alkaline hydrothermal vent," *Philosophical Transactions of the Royal Society* 362 (2007), 1887–925; author phone and email interviews with Michael J. Russell, January 29, 2013.

29 *For life to really get going*: David Bielo, "The Origin of Oxygen in Earth's Atmosphere," *Scientific American* (August 19, 2009), http://www.scientificamerican.com/article/origin-of-oxygen-in-atmosphere/.

29 *These guys promoted photosynthesis*: Lane, *Life Ascending*, 60–69.

30 *Oxygen made the planet livable*: Fred Guterl, *The Fate of the Species: Why the Human Race May Cause Its Own Extinction and How We Can Stop It* (New York: Bloomsbury, 2012), 41–44.

30 *life had to wait about four billion years*: "Oxygen-Free Early Oceans Likely Delayed Rise of Life on Planet," University of California, Riverside (January 10, 2011), http://newsroom.ucr.edu/2520.

30 *The Burgess Shale, the famous quarry of Cambrian life*: Leakey, *The Sixth Extinction*, 13–37.

31 *Our first really good display of what nature was up to*: Stephen Jay Gould, *Wonderful Life: The Burgess Shale and the Nature of History* (New York: W. W. Norton, 1989), 53–70.

31 *soft body parts*: Leakey, *The Sixth Extinction*, 15–16.

32 *They were an odd bunch*: Smithsonian National Museum of Natural History, Paleobiology, Burgess shale website: (a) http://paleobiology.si.edu/burgess

/opabinia.html; (b) http://paleobiology.si.edu/burgess/amiskwia.html; (c) http://paleobiology.si.edu/burgess/anomalocaris.html.

33 *The development of vision*: Lane, *Life Ascending*, 172–205.

33 *Animal life has grown quite larger*: Author trip in early summer 2012 to Kenya and Tanzania as a guest of Leslea Hlusko at the University of California, Berkeley, and Jackson Njau at Indiana University at Olduvai Gorge in Tanzania. Ngorongoro Basin was the first stop.

34 *Ngorongoro Crater became*: UNESCO, Culture, World Heritage Centre, "Ngorongoro Conservation Area: Outstanding Universal Value," http://whc.unesco.org/en/list/39.

35 *Despite the government threat to shoot poachers on sight*: Jeffrey Gettleman, "Elephants Dying in Epic Frenzy as Ivory Fuels Wars and Profits," *New York Times*, September 8, 2012, http://www.nytimes.com/2012/09/04/world/africa/africas-elephants-are-being-slaughtered-in-poaching-frenzy.html.

36 *Males use their tusks to battle each other*: Weiner, *The Beak of the Finch*, 263.

3 THE GROUND BELOW THE THEORIES

37 Beagle *rowed up to the island of San Cristóbal*: Weiner, *The Beak of the Finch*, 21–23 and 354–81.

39 *When Darwin returned to England*: Ibid., 28–29.

40 *William and Henry Blanford*: Ted Nield, *Supercontinent: Ten Billion Years in the Life of Your Planet* (Cambridge, MA: Harvard University Press, 2007), 30–35.

40 *Britain's Captain Robert Scott*: Ibid., 64–67; Sian Flynn, "The Race to the South Pole," *BBC History in Depth*, March 3, 2011, http://www.bbc.co.uk/history/british/britain_wwone/race_pole_01.shtml.

41 *Scott's second in command*: The National Museum: Royal Navy (UK), "Biography: Captain Robert Scott," *Royal Naval Museum Library*, 2004.

41 *Alfred Wegener, a German geophysicist*: Nield, *Supercontinent*, 14.

42 *island biogeography*: Ben G. Holt, et al., "An Update of Wallace's Zoogeographic Regions of the World," *Science* 339, no. 6115 (January 4, 2013), 74–78; UC Berkeley, "Biogeography: Wallace and Wegener," Understanding Evolution, http://evolution.berkeley.edu/evolibrary/article/history_16.

42 *considered an island for one organism*: Paul R. Ehrlich, David S. Dobkin, and Darryl Wheye, "Island Biogeography," 1988, https://www.stanford.edu/group/stanford birds/text/essays/Island_Biogeography.html.

43 *typical example of a landlocked island*: Michael Tennesen, "Expedition to the clouds," *International Wildlife* 28, no. 2 (March/April 1998), 22–29.

45 *all the birds mentioned in Shakespeare's*: Alan Weisman, *The World Without Us* (New York: Thomas Dunne Books/St. Martin's Press, 2007), 35.

45 *The brown tree snake*: Ker Than, "Drug-filled Mice Airdropped Over Guam to Kill Snakes," *National Geographic News*, September 24, 2010.

45 *Burmese pythons into Florida*: Michael Tennesen, "Python Predation: Big snakes poised to change US ecosystems," *Scientific American*, January 20, 2010.

46 *into the underground caverns of Powell's Cave*: Michael Tennesen, "When Juniper and Woody Plants Invade, Water May Retreat," *Science* 322, no. 5908 (December 12, 2008), 1630–31.

47 *thickets that don't allow enough light*: Steve Archer, David S. Schimel, and Elizabeth A. Holland, "Mechanisms of Shrubland Expansion: Land Use, Climate of CO2," *Climatic Change* 29 (1995), 91–99.

48 *Tallgrass Prairie Preserve near*: The Nature Conservancy, "Oklahoma Tallgrass Prairie Preserve," http://www.nature.org/ourinitiatives/regions/northamerica/unitedstates/oklahoma/placesweprotect/tallgrass-prairie-preserve.xml; I attended a conference session on this at the 2011 Ecological Society of American Convention in Austin, Texas.

49 *leaks in Boston's underground pipelines*: Nathan Phillips, et al., "Mapping urban methane pipeline leaks: methane leaks across Boston," *Environmental Pollution* 173 (2013), 1–4.

4 EVOLVING OUR WAY TOWARD ANOTHER SPECIES

50 *how man developed*: Tim D. White, "Human Evolution: The Evidence," in *Intelligent Thought: Science versus the Intelligent Design Movement*, edited by John Brockman (New York: Vintage, 2006), 65–81.

52 *studying their teeth*: Theresa M. Grieco, et al., "A Modular Framework Characterizes Micro- and Macroevolution of Old World Monkey Dentitions," *Evolution* 67, no. 1 (January 2013), 241–59.

53 *crocodiles and their possible effect on hominid intelligence*: Jackson K. Njau and Robert Blumenschine, "Crocodylian and mammalian carnivore feeding traces on hominid fossils from FLK 22 and FLK NN 3, Plio-Pleistocene Olduvai Gorge, Tanzania," *Journal of Human Evolution* 63, no. 2 (August 2012), 408–17.

54 *crocodile victims would show fewer tooth marks*: Jackson K. Njau and Robert Blumenschine, "A diagnosis of crocodile feeding traces on larger mammal bone, with fossil examples from Plio-Pleistocene Olduvai Basin, Tanzania," *Journal of Human Evolution* 50, no. 2 (2006), 142–62.

55 *In order to survive like this*: Author interviews with Jackson Njau and Leslea Hlusko, June 22–30, 2012.

56 Proconsul africanus: Douglas Palmer, *Origins: Human Evolution Revealed* (London: Mitchell Beazley, 2010), 35.

56 Australopithecus afarensis: Ibid., 58.

57 Homo habilis: Ibid., 98.

57 Homo erectus: Ibid., 119.

57 Homo sapiens: Ibid., 174–76.

58 *tool technology of early man:* Stephen S. Hall, "Last of the Neanderthals," *National Geographic* 214, no. 4 (October 2008), 38–59.

58 *Examinations of Neanderthal*: Palmer, *Origins*, 240.

58 *a day in the Tour de France*: Matt Allyn, "Eating for the Tour de France," *Bicycling*

Magazine, July 13, 2011, http://www.bicycling.com/garmin-insider/featured-stories/eating-tour-de-france.

59 Homo sapiens *moved up into Europe*: Palmer, *Origins*, 241.

59 "Homo sapiens *had the ability to develop trade*": Author interview with Rick Potts, May 3, 2012.

60 *"Mama," "papa," "cup," and "up"*: Author interview with Rob Shoemaker, November 18, 2011.

61 *if Bonnie had the ability to make decisions*: Chikako Suda-King, "Do orangutans (*Pongo pygmaeus*) know when they do not remember?" *Animal Cognition* 11, no. 1 (2008), 21–42.

62 *sign language and a system of lexicons*: Paul Raffaele, "Speaking Bonobo: Bonobos have an impressive vocabulary, especially when it comes to snacks," *Smithsonian* 37, no. 8 (November 2006), 74; author interview with Sue-Savage Rumbaugh, November 21, 2011.

62 *cognitive abilities to perceive speech*: Author interview with Lisa Heimbauer, November 17, 2011.

62 *the FOXP2 speech gene*: Elizabeth Kolbert, "Sleeping with the Enemy: What happened between the Neanderthals and us?" *The New Yorker* 87, no. 24 (August 15–22, 2011), 64–75.

63 *The town was established in 1781*: University of California, Los Angeles, "A Short History of Los Angeles," http://cogweb.ucla.edu/Chumash/LosAngeles.html.

64 *developed in the last one hundred years*: "Los Angeles County, California: Quick Facts from the UA Census Bureau," http://quickfacts.census.gov/qfd/states/06/06037.html.

64 *world's population growth*: Robert Kunzig, "Seven Billion: Special Series," *National Geographic*, January 2011.

65 *India's population growth*: Kenneth R. Weiss, "Beyond 7 Billion, Part 1: The Biggest Generation," *Los Angeles Times*, July 22, 2012.

66 *China's one-child policy*: Kenneth R. Weiss, "Beyond 7 Billion, Part 4: The China Effect," *Los Angeles Times*, July 22, 2012.

67 The Population Bomb: Paul Ehrlich and Anne Ehrlich, "The Population Bomb Revisited," *The Electronic Journal of Sustainable Development* 1, no. 3 (2009), Population-Bomb-Revisted-Paul-20096-5.pdf.

5 WARNING SIGN I: THE SOIL

71 *This land was transformed into grasslands*: Jan Zalasiewicz, *The Earth After Us: What Legacy Will Humans Leave in the Rocks?* (Oxford, UK: Oxford University Press, 2008), 86.

72 *Lawes, entrepreneur and agricultural scientist*: Rothamsted Research, "Rothamsted Research: Where Knowledge Grows," *Science Strategy*, 2012–17, www.rothamsted.ac.uk.

73 *reduction of biodiversity*: Weisman, *The World Without Us*, 192–94.

74 *"sample archive"*: Ibid., 194–202.

74 *tending plants was easier than hunting game*: Gregory Cochran and Henry Harpending, *The 10,000 Year Explosion: How Civilization Accelerated Human Evolution* (New York: Basic Books, 2009), 67–71.

76 *While we were nomadic*: Author interview with Ian Tattersall, curator emeritus of the American Museum of Natural History, April 18, 2012.

76 *sample archive*: Weisman, *The World Without Us*, 194–202.

77 *That afternoon, out in the fields*: Justin Gillis, "Norman Borlaug, Plant Scientist Who Fought Famine, Dies at 95," *New York Times*, September 13, 2009.

79 *20:20 Wheat*: "Rothamsted Research," *Science Strategy*, 2012–17, www.rothamsted.ac.uk.

79 *Jonathan Lynch*: http://plantscience.psu.edu/directory/jpl4; http://plantscience.psu.edu/research/labs/roots/about; author interview with Jonathan Lynch, August 2011.

79 *Susan McCouch*: http://vivo.cornell.edu/display/individual138; author interview with Susan McCouch, August 2011.

80 *potential of* terra preta: Michael Tennesen, "Black Gold of the Amazon: Fertile, charred soil created by pre-Columbian peoples sustained late settlements in the rain forest," *Discover Magazine* 28, no. 4 (April 2007), 46–52.

81 *Amazonian soils*: Manuel Arroyo-Kalin, "Slash-burn-and-churn," *Quaternary International* 249 (August 18, 2011), 4–18.

82 *to the Calhoun Experimental Forest*: Daniel Richter, et al., "Evolution of Soil, Ecosystem, and Critical Zone Research at the USDA FS Calhoun Experimental Forest," in *USDA Forest Service Experimental Forests and Ranges: Research for the Long Term* (New York: Springer, 2014).

83 *where the soil had been excavated*: Daniel Richter and Dan H. Yaalon, "The Changing Model of Soil, Revisited," *Soil Science Society of America Journal* 76, no. 3 (May 2012), 766–78.

83 *West-central Florida produces much of the US phosphorus*: Michael Tennesen, "Phosphorus Fields: Phosphorus and nitrogen fertilizers drive modern agriculture, but they are also poisoning the planet," *Discover*, December 2009, 55–59.

84 *Gaseous emissions of nitrogen can drift with the wind*: Michael Tennesen, "Sour Showers: Acid Rain Returns—This Time It Is Caused by Nitrogen Emissions," *Scientific American* 303, no. 3 (June 21, 2010), 23–24.

85 *sample from the year 1963*: John F. Kennedy Presidential Library and Museum, "Nuclear Test Ban Treaty," www.jfklibrary.org.

86 *green roofs*: Stuart R. Gaffin, Cynthia Rosenzweig, and Angela Y. Y. Kong, "Adapting to climate change through urban green infrastructure," *Nature Climate Change* 2, no. 704 (2012).

86 *author of* The Vertical Farm: Dickson Despommier, *The Vertical Farm: Feeding the World in the 21st Century* (New York: Thomas Dunne Books/St. Martin's Press, 2010), 3–11.

87 *cattle-rearing*: UN News Centre, "Rearing cattle produces more greenhouse gases than driving cars, UN report warns," November 29, 2006.

87 *soils are critical components of the earth's biosphere*: Ronald Almundson, "Protecting Endangered Soils," *Geotimes* 43, no. 3 (March 1998).

88 *43 percent of earth's land to agricultural production*: Author interview with Anthony Barnosky, March 2, 2012.

88 *murder of 800,000 Rwandans*: Robert Kunzig, "Seven Billion: Special Series," *National Geographic*, January 2011, 62.

6 WARNING SIGN II: OUR BODIES

89 *the story of Mr. Yu. G.*: Richard Preston, *The Hot Zone* (New York: Random House, 1994), 72–77; WHO Study Team, "Ebola haemorrhagic fever in Sudan, 1976," *Bulletin of the World Health Organization* 55, no. 2 (1978), 247–70, www.ncbi.nlm.nih.gov/pmc/articles/PMC2395561/.

90 *It then went through the hospital*: R. C. Baron, et al., "Ebola virus disease in southern Sudan: hospital dissemination and intrafamilial spread," *Bulletin of the World Health Organization* 61 (1983), 997–1003, http://www.ncbi.nlm.nih.gov/pmc/articles/PMC2536233/.

91 *Some scientists believe*: Preston, *Hot Zone*, 49; Joshua Hammer, "The Hunt for Ebola," Smithsonian.com, November 2012. http://www.smithsonianmag.com/science-nature/the-hunt-for-ebola-81684905/?all&no-ist.

91 *60 percent of all infectious diseases*: Augustin Estrada-Pena, et al., "Effects of Environmental change on zoonotic disease risk: an ecological primer," *Trends in Parasitology* 30 (April 2014), 205–14.

91 *severe acute respiratory syndrome (SARS)*: "Severe acute respiratory syndrome (SARS)," PubMed Health, http://www.ncbi.nlm.nih.gov/pubmedhealth/PMH0004460/.

92 *the effects of the disease are spread out and diluted*: Richard S. Ostfeld, "Are predators good for your health? Evaluating evidence for top-down regulation of zoonotic disease reservoirs," *Frontiers in Ecology and the Environment* 2, no. 1 (2004), 13–20.

93 *drive out or kill the bats*: Author interviews with Richard Ostfeld, 2011–14.

93 *increased the presence of infectious disease*: Cochran and Harpending, *The 10,000 Year Explosion*, 159–67.

94 *people in close proximity for disease to spread*: Ibid., 155–59.

94 *immunity to malaria can come at great cost*: Carolyn Sayre, "What You Need to Know About Sickle Cell Disease," *New York Times*, June 29, 2011.

94 *According to the World Health Organization*: WHO, "Malaria: Fact Sheet No. 94," *Media Centre*, December 2013, http://www.who.int/mediacentre.

95 *fewer domesticated animals*: Chengfeng Qin and Ede Qin, "Review of Bird Flu: A Virus of Our Own Hatching," *Virology Journal* 4 (April 2007), 38; video interview, Birdflubook.com.

96 *Francisco de Orellana*: Charles Mann, *1491: New Revelations of the Americas Before Columbus* (New York: Knopf, 2005), 315–21.

97 *"Wherever the European has trod"*: Cochran and Harpending, *The 10,000 Year Explosion*, 167–69.

97 *With these potential killers held at bay*: The College of Physicians of Philadelphia, "The History of Vaccines: Yellow Fever," 2014, www.historyofvaccines.org.

98 *Powassan virus encephalitis*: Lori Quillen, "Black-legged Ticks Linked to Encephalitis in New York State," Cary Institute of Ecosystem Studies, July 15, 2013, http://rhinebeck.wordpress.com/2013/07/15/black-legged-ticks-linked-to-encephalitis-in-new-york-state/.

100 *Small mammals are much better at handing off infections*: Jesse Brunner, Shannon Duerr, Felicia Keesing, Mary Killilea, Holly Vuong, and Richard S. Ostfeld, "An experimental test of competition among mice, chipmunks, and squirrels in deciduous forest fragments," *PLOS One*, June 18, 2013.

101 *fragmented forests increase disease*: Ibid.

102 *Antibiotic resistance*: Robert S. Lawrence, "The Rise of Antibiotic Resistance: Consequences of FDA's Inaction," *The Atlantic*, January 23, 2012.

103 *antimicrobial chemicals used in personal-care products*: John Cronin, "Antibacterial soaps don't work, are bad for humans & the environment," *EarthDesk*, December 19, 2013, http://earthdesk.blogs.pace.edu/2013/12/19/antibacterial-soaps-dont-work/.

104 *Common afflictions like gonorrhea*: WHO, "Urgent action needed to prevent the spread of untreatable gonorrhoea," *Media Centre*, June 6, 2012, http://www.who.int/mediacentre/news/notes/2012/gonorrhoea_20120606/en/.

104 *currently resurging is tuberculosis*: Mayo Clinic, "Tuberculosis," January 26, 2011, http://www.mayoclinic.com/health/tuberculosis/DS00372.

105 *These chronic diseases have overtaken infectious diseases*: Lawrence, "The Rise of Antibiotic Resistance."

106 *"Some microorganisms are resistant to nearly everything"*: "WHO Director-General addresses an expert advisory group on antimicrobial resistance," Geneva, Switzerland, September 19, 2013, http://www.who.int/dg/speeches/2013/stag_amr_20130919/en/.

7 WARNING SIGN III: SQUID AND SPERM WHALES

108 *Santa Rosalía fishermen pursue Humboldt squid*: Michael Tennesen, "Humboldt Squid: Masters of Their Universe," *Wildlife Conservation Magazine*, February 2009.

109 *low-oxygen zones in the water, a result of climate change*: Lothar Stramma, "Expanding Oxygen-Minimum Zones in the Tropical Oceans," *Science* 320, no. 5876 (May 2, 2008), 655–58.

110 *Gilly wonders what the long-term effects*: Author interview with William Gilly, February 28, 2012.

110 *oxygen minimum zones that reached almost 0 percent*: Lothar Stramma, "Ocean oxygen minimum expansion and their biological impacts," *Deep Sea Research Part I: Oceanographic Research Papers* 57, no. 4 (April 2010), 587–95.

111 *Northern California's hake fishery*: F. Chan, et al., "Emergence of Anoxia in the California Current Large Marine Ecosystems," *Science* 319, no. 5865 (February 15, 2008), 920.

112 *"White Shark Café"*: Michael Tennesen, "Science Sleuths: The White Shark Café," *National Wildlife*, July/August 2011.

112 *a measurable decrease in oxygen*: Lothar Stramma, Sunke Schmidtko, Lisa A. Levin, and Gregory C. Johnson, "Mismatch between observed and modeled trends in dissolved upper-ocean oxygen over the last 50 yr," *Biogeosciences* 57, no. 4 (April 2010), 587–95.

113 *Oxygen deprivation was a major source of extinction*: Kenneth R. Weiss, "Oxygen-poor ocean zones are growing," *Los Angeles Times*, May 2, 2008, http://www.latimes.com/nation/la-na-deadzone2-2008may02-story.html.

114 *the world's drama in their rearview mirror*: John Steinbeck and Edward Ricketts, *The Log from the Sea of Cortez* (New York: Viking Press, 1951), 4.

115 *trolled a couple of lines off the back of their boat*: Ibid., 76.

116 *The squid must have migrated*: Raphael D. Sagarin, "Remembering the Gulf: changes in the marine communities of the *Sea of Cortez* since the Steinbeck and Ricketts expedition of 1940," *Frontiers in Ecology* (2008), http://www.esajournals.org/doi/abs/10.1890/070067.

116 *Humboldt squid are also famously cannibalistic*: Unai Markaida, William F. Gilly, César A. Salinas-Zavala, Rigoberto Rosas-Luis, and J. Ashley T. Booth, "Food and Feeding of Jumbo Squid *Dosidicus Gigas* in the Central Gulf of California during 2005–2007," *California Cooperative Oceanic Fisheries Investigations Reports* 49 (2008).

118 *encountered "huge" conches and whelks*: Sagarin, "Remembering the Gulf."

119 *a greatly changed community of open-ocean fish*: William Gilly, "Searching for the Spirits of the *Sea of Cortez*," *Steinbeck Studies* 15, no. 2 (Fall 2004), 5–14.

119 *when he got to San Pedro Mártir Island*: Ibid.

120 *low-oxygen water from off the North Pacific*: Chan, et al., "Emergence of Anoxia in the California Current Large Marine Ecosystems."

122 *thought animals might dive to 325 to 650 feet*: Michael Tennesen, "Deep Sea Divers: How low can marine animals go?" *Wildlife Conservation*, June 2005.

122 *an adaptable breath-holding animal*: Michael Tennesen, "Testing the Depth of Life: Northern elephant seals migrate farther than any other mammal," *National Wildlife*, February/March 1999.

124 *power-dive to depths of up to one mile*: Julia S. Stewart, William F. Gilly, John C. Field, and John C. Payne, "Onshore-offshore movement of jumbo squid (*Dosidicus gigas*) on the continental shelf," *Deep Sea Research Part II: Topical Studies in Oceanography* 95 (October 15, 2013), 193–96.

125 *impacts to fish most commonly found on menus*: Gaia Vince, "How the world's oceans could be running out of fish," *BBC Future*, September 21, 2012, http://www.bbc.com/future/story/20120920-are-we-running-out-of-fish.

8 THE END

130 *New York's system of waterways*: Gretchen Daily and Katherine Ellison, *The New Economy of Nature: The Quest to Make Conservation Profitable* (Washington, DC: Island Press, 2002), 63–85.

133 *That's a lot of important functions*: Gretchen Daily, *Nature's Services: Societal Dependence on Natural Ecosystems* (Washington, DC: Island Press, 1997), 4.

133 *fifty-two thousand animal and plant species*: Bill Marsh, "Are We in the Midst of a Sixth Mass Extinction? A Tally of Life Under Threat," *New York Times*, Sunday Review, Opinion Pages, June 1, 2012.

134 *Communities relying on local goods*: Daily, *Nature's Services*, 295–99.

135 *two anticancer drugs*: Ibid., 263.

136 *studying jaguar movements in the tropical forest*: Michael Tennesen, "Room for the Jaguar?" *dukenvironment magazine*, Nicholas School of the Environment and Earth Sciences, Duke University (Fall 2006 Honor Roll Issue), 28–29.

138 *The habitat of the jaguar*: Dalia A. Conde, "Modeling male and female habitat difference for jaguar conservation," *Biological Conservation* 143, no. 9 (May 31, 2010), 1980–88.

140 *Guatemala has lost two-thirds of its original forested area*: "Guatemala's national forest programme—integrating agendas from the country's diverse forest regions," *National Forest Programme of Guatemala*, 2006.

140 *Hurricane Mitch hit Central America in 1998*: "Impact of Deforestation—1998: Hurricane Mitch," MongaBay.com, http://rainforests.mongabay.com/09mitch.htm.

141 *Guatemala has lost about 65,500 acres*: Danilo Valladares, "Guatemala: Relentless Devastation of Mangroves," Inter Press Service, July 16, 2009, http://www.ipsnews.net/2009/07/guatemala-relentless-devastation-of-mangroves/.

142 *Marshall, who grew up in Australia*: Author interview with Charles Marshall, April 4, 2012.

142 *Las Vegas is Spanish for "the Meadows"*: Michele Ferrari and Steven Ives, *Las Vegas: An Unconventional History* (New York: Bulfinch Press, 2005), 1–127.

144 *nature is the real treasure here*: Fernando Maestre, et al., "Plant Species Richness and Ecosystem Multifunctionality in Global Drylands," *Science* 335, no. 6065 (January 13, 2012), 214–18.

144 *A well-developed biological crust*: Michael Tennesen, "Turning to Dust: Around the globe, grasslands are turning to desert and free-flowing bits of dirt and rock are remaking the environment," *Discover Magazine*, May 2010.

146 *the Hoover Dam near Las Vegas*: Emma Rosi-Marshall, "Colorado River can be revived," *Poughkeepsie Journal*, September 11, 2011.

146 *all-important food sources for native fish*: Lori Quillen, "Dams Destabilize River Food Webs: Lessons from the Grand Canyon," Cary Institute for Ecosystem Studies, August 20, 2013.

146 *altered the river's ecosystem*: Wyatt F. Cross, et al., "Food-web dynamics in a large river discontinuum," *Ecological Monographs* 83, no. 3 (August 2013).

147 *the river is rapidly losing water*: Sally Deneen, "Feds Slash Colorado River Release to Historic Lows," *National Geographic*, August 16, 2013.

148 *lower than the US Dust Bowl Era*: Tennesen, "Turning to Dust," *Discover*, May 2010.

148 *not enough water for the city to survive*: "A majority on Earth face severe self-inflicted water woes within 2 generations," *AAAS and EurekAlert!* Water in Anthropocene Conference, Bonn, Germany, May 24, 2013.

149 *There's no place else for the birds to go*: Forest Isbell, et al., "High plant diversity is needed to maintain ecosystem services," *Nature* 477 (September 8, 2011), 199–202.

150 *Las Vegas could get there, too*: Bruce Babbitt, "Age-Old Challenge: Water and the West," *National Geographic*, June 1991, 2–3.

9 THE LONG RENEWAL

152 *recover from the Permian extinction*: Sarda Sahney and Michael J Benton, "Recovery from the most profound mass extinction of all time," *Proceedings of the Royal Society, Biological Sciences* 275, no. 1636 (April 7, 2008), 759–65.

152 *The explosion toppled most of the trees*: Lyn Garrity, "Evolution World Tour: Mount St. Helens, Washington—Over thirty years after the volcanic eruptions, plant and animal life has returned to the disaster site, a veritable living laboratory," *Smithsonian Magazine*, January 2012, http://www.smithsonianmag.com/evotourism/evolution-world-tour-mount-st-helens-washington-6011404/.

153 *most studied volcano in the world*: P. Frenzen, "Life Returns: Frequently Asked Questions about Plant and Animal Recovery Following the 1980 Eruption," US Forest Service, Mount St. Helens Volcanic Monument, http://www.fs.usda.gov/mountsthelens.

154 *When I visited the park thirty years later*: "Mount St. Helens, 30 Years Later: A Landscape Reconfigured," Pacific Northwest Research Station, http://www.fs.fed.us/pnw/mtsthelens/.

155 *explosion became international news at the speed of electronic transmission*: Simon Winchester, "Krakatoa, the first modern tsunami," BBC, January 8, 2005.

155 *explosive noises for almost two months*: Simon Winchester, *Krakatoa: The Day the World Exploded: August 27, 1883* (New York: HarperCollins, 2003), 149–76.

155 *what has happened to the area since the eruption?*: Ian Thornton, "Figs, frugivores and falcons: an aspect of the assembly of mixed tropical forest on the emergent volcanic island, Anak Krakatau," *South Australian Geographical Journal* 93 (1994), 3–21.

157 *wholesale species changes*: Erwin, *Extinction*, 218–19.

158 *Fray Jorge National Park*: Michael Tennesen, "The Strange Forests That Drink—and Eat—Fog," *Discover*, April 2009.

159 *Barren land free of all vegetation gradually began to disappear*: Erwin, *Extinction*, 221.

160 *dominant vertebrate animal in the early Triassic*: Ibid., 235.

161 *they weren't the second-best predator of the day*: Randall B. Irmis, Sterling J. Nesbitt, and Hans-Dieter Sues, "Early Crocodylomorpha," in *Anatomy, Phylogeny and Palaeobiology of Early Archosaurs and their Kin*, edited by Sterling J. Nesbitt, Julia Brenda Desojo, and Randall B. Irmis (London: Geological Society, Special Publications, 2013).

161 *A colossal phytosaur looking like a diesel truck*: Scott Wing and Hans-Dieter Sues, *Mesozoic and Early Cenozoic Terrestrial Ecosystems* (Chicago: The University of Chicago Press, 1992), xii.

163 *there were mountain lions from his study nearby*: Michael Tennesen, "Can the Military Clean Up Its Act?" *National Wildlife*, October 1993.

164 *The Korean demilitarized zone*: Tom O'Neil, "Korea's DMZ: Dangerous Divide," *National Geographic*, July 2003, http://ngm.nationalgeographic.com/features/world/asia/north-korea/dmz-text/1.

164 *the perfect wildlife sanctuary*: Tim Wall, "War of Peace May Doom Korean DMZ Wildlife," *Discovery News*, March 18, 2013, http://news.discovery.com/earth/what-would-a-new-korean-war-do-to-dmz-wildlife-130318.htm.

165 *reproductive rates are much lower in Chernobyl birds*: A. P. Moller, et al., "Condition, reproduction and survival of barn swallows from Chernobyl," *Journal of Animal Ecology* 74 (2005), 1102–11.

165 *brains of the local birds are 5 percent smaller*: University of South Carolina, "Researcher finds birds have smaller brains," February 10, 2011, http://www.sc.edu/news/newsarticle.php?nid=1562#.U-1S9kjbaot.

166 *radiation effects will diminish over time*: University of Portsmouth, "Wildlife thriving after nuclear disaster?" *ScienceDaily*, April 11, 2012; J. T. Smith, N. J. Willey, and J. T. Hancock, "Low dose ionizing radiation produces too few reactive oxygen species to directly affect antioxidant concentrations in cells," *Biology Letters* (April 11, 2012), http://rsbl.royalsocietypublishing.org/content/early/2012/04/05/rsbl.2012.0150.

10 TROUBLED SEAS: THE FUTURE OF THE OCEANS

167 *lowers the pH of ocean waters, which is bad for krill*: Michael Marshall, "Animals are already dissolving in Southern Ocean," *New Scientist*, November 25, 2012.

168 *decreases the ability of whales to hear the mating calls of others*: Yifei Wang, "A Cacophony in the Deep Blue: How Acidification May Be Deafening Whales," *Dartmouth Undergraduate Journal of Science*, February 22, 2009.

168 *a ring of male humpback whales*: Michael Tennesen, "Tuning in to Humpback Whales," *National Wildlife*, February/March 2002.

169 *ocean was doing a good job of taking in CO_2 all by itself*: Andrew Revkin, "Papers Find Mixed Impacts on Ocean Species from Rising CO_2," *New York Times*, August 26, 2013.

169 *Acidification of ocean water is bad for krill*: Australian Antarctic Division, "Krill face deadly cost of ocean acidification," *Media News*, October 13, 2010.

170 *As the ocean gets noisier, whale sounds may get muffled*: Keith C. Hester, Edward T. Peltzer, William J. Kirkwood, and Peter G. Brewer, "Unanticipated consequences of ocean acidification: A noisier ocean at lower pH," *Geophysical Research Letters* 35 (October 1, 2008), L19601.

170 *unable to get their gills back*: Author interview with Hans-Dieter Sues, April 16, 2012.

171 *"Atlantification" of the Arctic*: Curt Stager, *Deep Future: The Next 100,000 Years of Life on Earth* (New York: Thomas Dunne Books, 2011), 150.

172 *the worst thing about McMurdo is the food*: Author interview with Gretchen Hofmann, February 27, 1992.

172 *the bounds of natural pH fluctuation*: Gretchen E. Hofmann, et al., "High-Frequency Dynamics of Ocean pH: A Multi-Ecosystem Comparison," *PLOS One*, December 2011.

172 *Curt Stager, author of* Deep Future: Stager, *Deep Future*, 156–58.

173 *CO_2 vents off the island of Ischia*: J. Garrabou, et al., "Mass mortality in Northwestern Mediterranean rocky benthic communities," *Global Change Biology* 15, no. 5 (May 2009), 1090–103.

174 *"Elvis Taxa"*: Erwin, *Extinction*, 237.

174 *the loss of coral*: Rebecca Albright, Benjamin Mason, Margaret Miller, and Chris Langdon, "Ocean acidification compromises recruitment of the threatened Caribbean coral *Acropora palmata*," *Proceedings of the National Academy of Sciences* 107, no. 47 (2010), 20400–4.

175 *Acidification there dissolves the shells of sea snails*: Marshall, "Animals are already dissolving in Southern Ocean."

176 *The development of the frozen food industry*: Callum Roberts, *The Unnatural History of the Sea: The Past and Future of Humanity and Fishing* (London: Gaia, 2007), 201–2.

177 *armorhead fish around seamounts off Hawaii*: Ibid., 290.

177 *a full-scale assault on the fishery*: Ibid., 291.

178 *harvesting deep-sea sediments for their rare-earth metals*: Author interview with Craig McClain, April 11, 2012.

179 *"Jellyfish Gone Wild"*: National Science Foundation, "Environmental Change and Jellyfish Swarms," http://www.nsf.gov/news/special_reports/jellyfish/.

179 *one of their favorite haunts is Jellyfish Lake*: Pamela S. Turner, "Darwin's Jellyfishes," *National Wildlife*, August/September 2006.

179 *refrigerator-sized Nomura's jellyfish*: Blake de Pastino, "Giant Jellyfish Invade Japan," *National Geographic News*, October 28, 2010.

180 *slow-moving, drifting animals with no vision*: Jose Luis Acuña, "Faking Giants: The Evolution of High Prey Clearance Rates in Jellyfish," *Science* 333, no. 6049 (September 16, 2011), 1627–29.

180 *Jellyfish are ancient, dating back 600 million*: Anahad O'Connor, "No Fins? No Problem: Jellyfish Have Their Ways," *New York Times*, September 30, 2011.

180 *overfishing is promoting the presence of jellyfish*: Institut de Recherche pour le Développement, "Boom in jellyfish: Overfishing called into question," *Science-Daily*, May 3, 2013.

182 *his crew struggled to find calm waters*: Roberts, *The Unnatural History of the Sea*, xi–xii.

182 *Today, Palmyra Atoll, 1,000 miles south of Hawaii*: Nature Conservancy, "Palmyra: A Marine Wilderness: Palmyra is that rare place where top predators such as sharks still dominate the reef ecosystem," www.nature.org/ourinitiatives/regions/northamerica/unitedstates/hawaii/palmyraatoll/.

11 PREDATORS WILL SCRAMBLE

184 *the role of sea otters as predators*: J. A. Estes, D. F. Doak, A. M. Springer, and T. M. Williams, "Causes and consequences of marine mammal population declines in southwest Alaska," *Philosophical Transactions of the Royal Society* 364, no. 1524 (June 25, 2009), 1647–58.

185 *never dawned on me that that would be an interesting question*: Author interview with James Estes, March 1, 2012.

185 *kelp rises like an undersea forest*: Monterey Bay Aquarium, "Giant kelp, Natural History," http://www.montereybayaquarium.org/animal-guide/plants-and-algae/giant-kelp.

186 *no predatory pressure on the urchins*: J. A. Estes, M. T. Tinker, and J. L. Bodkin, "Using Ecological Function to Develop Recovery Criteria for Depleted Species: Sea Otters and Kelp Forests in the Aleutian Archipelago," *Conservation Biology* 24, no. 3 (June 2010), 1523–739.

187 *In 2011, Estes—along with John Terborgh*: James A. Estes, et al., "Trophic Downgrading of Planet Earth," *Science* 333, no. 6040 (July 15, 2011), 301–6.

188 *in the absence of predators, prey populations exploded*: John Terborgh, et al., "Ecological Meltdown in Predator-Free Forest Fragments," *Science* 294, no. 5548 (November 2001), 1923–26.

188 *predators played a major part in making the world green*: "The Lake Guri Experiment," *National Geographic's Strange Days on Planet Earth*, www.pbs.org/strangedays/episodes/predators/experts/lakeguri.html.

188 *Sharks are another vulnerable predator*: Dalhousie University, "Shark Fisheries Globally Unsustainable," *Newswise*, March 1, 2013.

189 *studying hammerhead shark populations*: Brian Handwerk, "Do Hammerheads Follow Magnetic Highways to Migration?" *National Geographic News*, June 6, 2002.

190 *Large tiger sharks can grow to 20 to 25 feet*: Michael Tennesen, "In Hawaii, scientists are helping to dispel some myths about tiger sharks," *National Wildlife*, August 2000.

190 *Bull sharks get their name*: Michael Tennesen, "The Bull Shark's Double Life," *National Wildlife*, November 2011.

191 *shark ecotourism currently generates more than $314 million a year*: Andres Cisneros-Montemayor, et al., "Global economic value of shark ecotourism," *Oryx*, July 2013.

192 *attaching tracking devices to elephant seals*: Tennesen, "Testing the Depth of Life," *National Wildlife*, February 1999.

193 *recorded their hunting in tightly coordinated groups*: Russ Vetter, "Interactions and niche overlap between mako shark, *Isurus oxyrinchus*, and Jumbo Squid," *California Cooperative Oceanic Fisheries Investigations Reports* 49 (2008).

193 *Giant squid are the biggest invertebrates*: "Giant Squid, *Architeuthis dux*," photo, *National Geographic*, http://animals.nationalgeographic.com/animals/invertebrates/giant-squid/.

194 *limbs equipped with sharp claws or hooks*: Museum of New Zealand, "Colossal Squid: Hooks and suckers," Te Papa Museum, April 30, 2008, http://squid.tepapa.govt.nz/anatomy/article/the-arms-and-tentacles.

195 *Cuttlefish also have a rich vocabulary*: Marguerite Holloway, "Cuttlefish Say It With Skin," *Natural History*, April 2000.

195 *octopuses are more intelligent than any of the fishes*: Michael Tennesen, "Outsmarting the Competition: When it comes to intelligence and personality, the giant Pacific octopus shines," *National Wildlife*, December 2002.

196 *They currently live to about one and a half years of age*: From "Chinese commercial fishery data off Costa Rica," provided by William Gilly.

12 THE DECLINE AND RETURN OF MEGAFAUNA

199 *But we're not creeping anymore*: Author interview with Charles Marshall, April 4, 2012.

200 *settlers apparently hunted giant moas for generations*: Leakey, *The Sixth Extinction*, 185.

201 *early hunters shut off all the methane*: Felisa A. Smith, Scott M. Elliot, and Kathleen Lyons, "Methane emissions for extinct megafauna," *Nature Geoscience* 3 (2010), 374–75.

201 *predators show heavier tooth wear*: Brian Switek, "Broken teeth tell of tough times for Smilodon," *ScienceBlogs*, February 15, 2010, http://scienceblogs.com/laelaps/tag/tar-pits/.

202 *Animals grew about eight orders of magnitude*: Alistair R. Evans, et al., "The maximum rate of mammal evolution," *Proceedings of the National Academy of Sciences*, October 1, 2011, http://www.pnas.org/content/early/2012/01/26/1120774109.abstract; National Science Foundation, "Mammals grew 1,000 times larger after the demise of the dinosaurs," NSF, November 25, 2010, http://www.nsf.gov/news/news_summ.jsp?cntn_id=118129.

203 *The Tibetan woolly rhino lived at a much warmer time*: Tao Deng, "Out of Tibet: Pliocene Wooly Rhino Suggests High-Plateau Origin of Ice Age Megaherbivores," *Science* 333, no. 6047 (September 2, 2011), 1285–88.

203 *it's known as Bergmann's Rule*: Melissa I. Pardi and Felisa A. Smith, "Paleoecology in an Era of Climate Change," in *Paleontology in Ecology and Conservation*, edited by J. Louys (Berlin: Springer-Verlag, 2012).

204 *an indication of pack rat size and diet*: Steve Carr, "UNM Researchers Explore Evolution of World's Mammals Over the Past 100 Million Years," University of New Mexico, November 25, 2010.

205 *had adapted to eating all meat, all of the time*: Blaire Van Valkenburgh, Xiaoming Wang, and John Damuth, "Cope's Rule, Hypercarnivory, and Extinction in North American Canids," *Science* 306, no. 5693 (October 1, 2004), 101–4.

205 *restore all the big animals that once stocked North America*: C. Josh Donlan, et al., "Pleistocene Rewilding: An Optimistic Vision for 21st Century Conservation," *The American Naturalist* 168, no. 5 (November 2006), 660–81.

206 *can't just reintroduce a large herbivore like a horse*: J. W. Turner, Jr. and M. Morrison, "Trends in number and survivorship of a feral horse population: Influence of mountain lion predation," *Southwestern Naturalist* (2001).

207 *The American cheetah fist appeared perhaps 2.5 million years ago*: Donlan, "Pleistocene Rewilding." Donlan's paper proposes and explains many of the animals mentioned in this section on rewilding.

211 *Wolves were introduced in the 1990s to Yellowstone National Park*: US National Park Service, "Wolf Restoration," http://www.nps.gov/yell/naturescience/wolfrest.htm.

211 *there were many more predators and prey*: Gary W. Roemer, Matthew W. Gompper, and Blaire Van Valkenburgh, "The Ecological Role of Mammalian Mesocarnivore," *BioScience* 59, no. 2 (February 2009), 165–73.

212 *working on the Reserva de Faia Brava in Portugal*: Staffan Widstrand, "Western Iberia II—Faia Brava, Portugal," Rewilding Europe, April 20, 2011, http://www.rewildingeurope.com/areas/western-iberia/.

213 *the best candidates for a future laid waste by man*: Ward, *Future Evolution*, 105.

214 *identified the canine tooth of a saber-toothed cat*: John M. Harris, "Bones from the Tar Pits: La Brea Continues to Bubble Over with New Clues," *Natural History*, June 2007.

215 *I met with UCLA paleontologist Blaire Van Valkenburgh*: Author interview with Blaire Van Valkenburgh, May 28, 2013.

215 *Modern species evolved with larger and much more complex and large predators*: Blaire Van Valkenburgh, "Tough Times in the Tar Pits," *Natural History*, April 1994.

216 *this city was a wide floodplain*: Harris, "Bones from the Tar Pits."

13 INVADERS TO MARS?

219 *I ventured into the observatory that Lowell built*: Michael Tennesen, "Stars in Their Eyes: The Exquisite Telescopes Crafted by Alvan Clark and His Sons Helped Make the Last Half of the 19th Century a Golden Age of Astronomy," *Smithsonian* 32, no. 7 (October 2001), 78–82.

220 *The fourth planet from the sun is more like Earth*: Michael Tennesen, "Mars: Remembrance of Life Past," *Discover*, July 1989.

221 *It would be an inhospitable place for a moonlit walk*: Robert Zubrin and Richard Wagner, *The Case for Mars: The Plan to Settle the Red Planet and Why We Must* (New York: Free Press/Simon and Schuster, 1996), xxiv.

221 *the* Curiosity *rover in 2013 analyzed a powdered sample*: Jet Propulsion Laboratory, "Mars Rover Carries Device for Underground Scouting," October 20, 2011, http://www.jpl.nasa.gov/news/news.php?release=2011-325.

223 *enough gas to make the round trip*: Zubrin and Wagner, *The Case for Mars*, 161–70.

223 *Zubrin agrees but thinks that a small crew would still be best*: Ibid., 98–99.

224 *Geothermal power would be an attractive source*: Ibid., 226–27.

224 *the possibility of interstellar trade*: Ibid., 242–43.

225 *The first Martian task would be to find water*: Ibid., 211–16.

226 *Mars might also be a good place to understand our past*: Amina Khan, "Study: Mars could have held watery underground oases for life," *Los Angeles Times*, January 21, 2013.

226 *all life on Earth is descended from microorganisms on Mars*: Rick Fienberg, "MIT: Are you a Martian? We could all be, scientists say," *American Astronomical Society*, March 23, 2011.

227 *The ice provides what scientists call "thermal buffering"*: Tennesen, "Mars."

228 *Using CF gases [as opposed to CFCs]*: Author interview with Robert Zubrin, October 1, 2013.

228 *space travel may have to be an international effort*: Chris Bergin, "Fobos-Grunt ends its misery via re-entry," NASA, January 15, 2012, http://www.nasaspaceflight.com/2012/01/fobus-grunt-ends-its-misery-via-re-entry/.

230 *a DNA sequencing machine that could be landed on the surface of Mars*: Antonio Regalado, "Genome Hunters Go After Martian DNA," *Technology Review*, October 18, 2012.

230 *You would live your life out on Mars*: BBC, "How to get along for 500 days together," *BBC News Magazine*, March 1, 2013.

230 *Lots to offer here, just no welcome-home party*: Alan Boyle, "One-way Mars trip attracts 165,000 would-be astronauts . . . and counting," NBC News, August 23, 2013, http://www.nbcnews.com/science/space/one-way-mars-trip-attracts-165-000-would-be-astronauts-f6C10981032.

231 *Biosphere 2 offers a way to study*: Author interview with John Adams, November 5, 2012.

231 *dependent on the facility's different biomes and infrastructure*: J. P. Allen, M. Nelson, and A. Alling, "The legacy of Biosphere 2 for the study of biospherics and closed ecological systems," *Advances in Space Research* 31, no. 7 (2003), 1629–39.

231 *they also needed to adjust to the levels of* CO_2 *that fluctuated wildly*: Abigail Alling, Mark Nelson, and Sally Silverstone, *Life under Gas: The Inside Story of Biosphere 2* (Oracle, AZ: The Biosphere Press, 1993).

233 *They develop bird legs*: National Space Biomedical Research Institute (NSBRI), "How the Human Body Changes in Space," http://www.nsbri.org/DISCOVERIES-FOR-SPACE-and-EARTH/The-Body-in-Space/.

234 *Is it someplace where large portions of our population might escape*: Richard Hollingham, "Building a new society in space," BBC Future, March 18, 2013, http://www.bbc.com/future/story/20130318-building-a-new-society-in-space.

235 *habitable planets like Earth around other stars*: Dennis Overbye, "2 Good Places to Live, 1,200 Light-Years Away," *New York Times*, April 18, 2013.

14 IS HUMAN EVOLUTION DEAD?

236 *kill large animals without having to have large muscles*: Cochran and Harpending, *The 10,000 Year Explosion*, 3–5.

237 *one hundred times faster than the long-term average*: Ibid., back cover.

238 *Neither of them looks like a wolf*: Ibid., 6.

239 *differences between the Viking invaders and their peaceful Swedish descendants*: John Hawks, Eric T. Wang, Gregory M. Cochran, Henry C. Harpending, and Robert K. Moyzis, "Recent acceleration of human adaptive evolution," *Proceedings of the National Academy of Sciences* 104, no. 52 (December 26, 2007), 20753–58.

239 Homo sapiens *migrated into Eurasia*: Guy Gugliotta, "The Great Human Migration: Why humans left their homeland 80,000 years ago to colonize the world," *Smithsonian Magazine*, July 2008, http://www.smithsonianmag.com/history/the-great-human-migration-13561/.

240 *They're different. But you can't see it*: Weiner, *The Beak of the Finch*, 126.

240 *game-changing mutations to the human genome was lactose tolerance*: Cochran and Harpending, *The 10,000 Year Explosion*, 173–86.

241 *I got to witness this up close*: Michael Benanav, "Through the Eyes of the Maasai," *New York Times*, August 9, 2013.

243 Homo sapiens *were once pushed to the edge of extinction*: Robert Krulwich, "How Human Beings Almost Vanished from Earth in 70,000 BC," *Robert Krulwich on Science*, NPR, October 22, 2012, http://www.npr.org/blogs/krulwich/2012/10/22/163397584/how-human-beings-almost-vanished-from-earth-in-70-000-b-c.

244 *large-bodied herbivorous dinosaurs were declining*: Stephen L. Brusatte, Richard J. Butler, Albert Prieto-Marquez, Mark A. Norell, et al., "Dinosaur Morphological Diversity and the End-Cretaceous Extinction," *Nature Communications* 3, no. 804 (May 1, 2012).

245 *a series of glacial cycles that have warmed and cooled the earth*: Pardi and Smith, "Paleoecology in an Era of Climate Change."

245 *Highly resolved ice cores from Greenland*: European Geoscience Union, "The Oldest Ice Core—Finding a 1.5 Million-year Record of Earth's Climate," November 5, 2013, http://www.egu.eu/news/77/the-oldest-ice-core-finding-a-15-million-year-record-of-earths-climate/.

246 *The end of the Younger Dryas*: NOAA Paleoclimatology Program, "A Paleo Perspective on Abrupt Climate Change," August 20, 2008, www.ncdc.noaa.gov/paleo/abrupt/data4.html.

246 *Rhinos ran through the British brush*: Stager, *Deep Future*, 62.

247 *The Arctic will be one of the first areas to go*: National Snow and Ice Data Center, "Quick Facts on Arctic Sea Ice," 2013, http://nsidc.org/cryosphere/quickfacts /seaice.html.

247 *An Arctic passage from Rotterdam to Seattle*: Stager, *Deep Future*, 158.

248 *Summer Arctic sea ice*: Brad Plumer, "Arctic sea ice just hit a record low," *Washington Post*, August 28, 2012.

250 *global warming may actually put off the next ice age by thousands of years*: Christine Dell'Amore, "Next Ice Age Delayed by Global Warming, Study Says," *National Geographic News*, September 3, 2009.

15 BEYOND *HOMO SAPIENS*

251 *something like twenty-five human species; why couldn't there be another*: Juan Enriquez, "Will our kids be a different species?" TED Talk, April 2012, https://www.ted.com/talks/juan_enriquez_will_our_kids_be_a_different_species.

252 *there may have been four different species on the planet*: Nicholas Wade, "Genetic Data and Fossil Evidence Tell Differing Tales of Human Origins," *New York Times*, July 26, 2012.

253 *Americans in general have been growing taller, living longer, and getting thicker*: Patricia Cohen, "Technology Advances; Humans Supersize," *New York Times*, April 26, 2011.

253 *Allopatric speciation*: Author interview with Scott Carroll, July 1, 2011.

254 *studied populations of a small salamander,* Ensatina, *in the 1950s*: R. C. Stebbins, "Speciation in salamanders of the plethodontid genus *Ensatina*," *University of California Publications in Zoology*, 1949, http://evolution.berkeley.edu/evolibrary /article/0_0_0/devitt_02.

254 *intermarriage with non-Jews, as well as conversions to Judaism, were quite rare*: Cochran and Harpending, *The 10,000 Year Explosion*, 220.

256 *Googleplex buildings have high ceilings*: Julie Bort, "Tour Google's Luxurious 'Googleplex' Campus in California," *Business Insider*, October 6, 2013, http://www.businessinsider.com/google-hq-office-tour-2013-10?op=1.

257 *drawn to others of their kind, which could lead to speciation*: Peter Ward, "What will become of *Homo sapiens*?" *Scientific American*, January 2009.

257 *born from one of its mother's skin cells*: "Xiao Xiao Receives Torch Lit by Dolly," *China Daily*, August 8, 2009, www.chinadaily.com.cn.

258 *attempting to map the human brain*: Anne Trafton, "Illuminating neuron activity in 3-D," *MIT News*, May 18, 2014.

258 *the possibilities involved in uploading one's mind*: Nick Bostrom, "The Future of Human Evolution," in *Death and Anti-Death: Two Hundred Years After Kant, Fifty Years After Turing*, edited by Charles Tandy (Palo Alto, CA: Ria University Press, 2004), 339–71, http://www.nickbostrom.com/fut/evolution.html.

259 *Second Life is a 3-D online community*: Michael Tennesen, "Avatar Acts: When the

Matrix has you, what laws apply to settle conflicts?" *Scientific American* 301 (July 2009), 27–28.

260 *national defense institutions are among the most active investors in AI*: James Barrat, *Our Final Invention: Artificial Intelligence and the End of the Human Era* (New York: Thomas Dunne Books/St. Martin's Press, 2013), 25, 171–72.

262 *Georgii Gause pondered the options*: Charles C. Mann, "State of the Species: Does success spell doom for *Homo sapiens*?" *Orion Magazine*, November/December 2012, http://www.orionmagazine.org/index.php/articles/article/7146.

263 *gave a name to our time: the "Anthropocene"*: Andrew Revkin, "Confronting the 'Anthropocene,'" *New York Times*, May 11, 2011.

263 *"Then all bets are off"*: Author interview with Ian Tattersall, April 18, 2012.

264 *take a Neanderthal, clean him up, give him a haircut*: Palmer, *Origins*, 138.

265 *if our planet's beginning is the end of your nose*: Peter Ward and Donald Brownlee, *The Life and Death of Planet Earth: How the New Science of Astrobiology Charts the Ultimate Fate of Our World* (New York: Times Books, 2003), 14.

265 *4.5-billion-year history of our planet in terms of a twenty-four-hour day*: Northern Arizona University, "The History of Life on Earth: The 24-Hour Clock Analogy," http://www2.nau.edu/~lrm22/lessons/timeline/24_hours.html.

266 *The more specimens, the more chance for fossils*: Zalasiewicz, *The Earth After Us*, 120–21, 198.

267 *moved its Doomsday Clock forward to five minutes to midnight*: "Doomsday Clock Moves One Minute Closer to Midnight," *The Bulletin of the Atomic Scientists*, January 10, 2012, http://thebulletin.org/timeline.

269 *all continue to try to overrun their environments*: Mann, "State of the Species."

269 *But in a few decades in the nineteenth century, slavery almost vanished*: Ibid.

270 *Life may exist on earth, but it will be microscopic*: Ward, *Future Evolution*, 175.

271 *large reptile herbivores mired in a long-term decline*: "Were Dinosaurs Undergoing Long-Term Decline Before Mass Extinction?" American Museum of Natural History, October 26, 2012, http://www.amnh.org/our-research/science-news2/2012/were-dinosaurs-undergoing-long-term-decline-before-mass-extinction.

Selected Sources

BOOKS

Alley, Richard B. *The Two-Mile Time Machine: Ice Cores, Abrupt Climate Change, and Our Future*. Princeton, NJ: Princeton University Press, 2000.

Alling, Abigail, and Mark Nelson. *Life Under Glass: The Inside Story of Biosphere 2.* Oracle, AZ: Biosphere Press, 1993.

Alonso, L. E., A. Alonso, T. S. Schulenberg, and F. Dallmeier, eds. *Biological and Social Assessments of the Cordillera de Vilcabamba, Peru* (Washington, DC: Conservation International, Center for Applied Biodiversity Sciences, 2001).

Barnosky, Anthony D. *Heatstroke: Nature in an Age of Global Warming*. Washington, DC: Island Press, Shearwater Books, 2009.

Barrat, James. *Our Final Invention: Artificial Intelligence and the End of the Human Era*. New York: Thomas Dunne/St. Martin's Press, 2013.

Behrensmeyer, Anna K., John D. Damuth, William A. DiMichele, Richard Potts, Hans-Dieter Sues, and Scott L. Wing. *Terrestrial Ecosystems Through Time: Evolutionary Paleoecology of Terrestrial Plants and Animals*. Chicago: University of Chicago Press, 1992.

Cochran, Gregory, and Henry Harpending. *The 10,000 Year Explosion: How Civilization Accelerated Human Evolution*. New York: Basic Books, 2009.

Conniff, Richard. *The Species Seekers: Heroes, Fools, and the Mad Pursuit of Life on Earth*. New York: W. W. Norton & Company, 2011.

Daily, Gretchen C. *Nature's Services: Societal Dependence on Natural Ecosystems*. Washington, DC: Island Press, 1997.

Daily, Gretchen C., and Katherine Ellison. *The New Economy of Nature: The Quest to Make Conservation Profitable*. Washington, DC: Island Press/Shearwater Books, 2002.

Darwin, Charles. *On the Origin of Species by Means of Natural Selection*. New York: Bantam Books, 1999. First published 1859; still the premier text of evolutionary biology.

———. *The Voyage of the Beagle*. Ware, Hertfordshire, UK: Wordsworth Editions, 1997. First published in 1839.

Dawkins, Richard. *The Ancestor's Tale: A Pilgrimage to the Dawn of Evolution*. Boston: Houghton Mifflin, 2004.

———. *The Greatest Show on Earth: The Evidence for Evolution*. New York: Free Press, 2009.

Despommier, Dickson. *The Vertical Farm: Feeding the World in the 21st Century*. New York: Thomas Dunne Books/St. Martin's Press, 2010. Turning abandoned skyscrapers into future farms.

Diamond, Jared. *Guns, Germs, and Steel: The Fates of Human Societies*. New York: W. W. Norton, 1998.

———. *The Third Chimpanzee: The Evolution and Future of the Human Animal.* New York: HarperCollins, 1992.

Erwin, Douglas. *Extinction: How Life on Earth Nearly Ended 250 Million Years Ago.* Princeton, NJ: Princeton University Press, 2006. The classic book on the Permian extinction.

Ferrari, Michele, and Steven Ives. *Las Vegas: An Unconventional History.* New York: Bulfinch Press, 2005.

Finlayson, Clive. *The Humans Who Went Extinct: Why Neanderthals Died Out and We Survived.* New York: Oxford University Press, 2009.

Gould, Stephen Jay. *Ever Since Darwin: Reflections on Natural History.* New York: W. W. Norton, 1977.

———. *The Structure of Evolutionary Theory*. Cambridge, MA: Belknap Press of Harvard University Press, 2002. Gould's 1,343-page opus on punctuated equilibrium and his other thoughts on Darwinian evolution.

———. *Wonderful Life: The Burgess Shale and the Nature of History*. New York: W. W. Norton, 1989.

Guterl, Fred. *The Fate of the Species: Why the Human Race May Cause Its Own Extinction and How We Can Stop It*. New York: Bloomsbury, 2012.

Jackson, Rob. *The Earth Remains Forever: Generations at a Crossroads*. Austin: University of Texas Press, 2002.

Jones, Steve. *The Darwin Archipelago: The Naturalist's Career Beyond "Origin of Species."* New Haven, CT: Yale University Press, 2011.

Kolbert, Elizabeth. *The Sixth Extinction: An Unnatural History*. New York: Henry Holt, 2014.

Lane, Nick. *Life Ascending: The Ten Great Inventions of Evolution*. New York: W. W. Norton, 2009.

Leakey, Richard, and Roger Lewin. *The Sixth Extinction: Patterns of Life and the Future of Humankind.* New York: Doubleday, 1995.

Mann, Charles C. *1491: New Revelations of the Americas Before Columbus*. New York: Alfred A. Knopf, 2005.

Martin, Paul S. *Twilight of the Mammoths: Ice Age Extinctions and the Rewilding of America*. Berkeley: University of California Press, 2005.

National Research Council, Committee on Twenty-first Century Systems Agriculture. *Toward Sustainable Agricultural Systems in the 21st Century*. Washington, DC: National Academies Press, 2010.

Nield, Ted. *Supercontinent: Ten Billion Years in the Life of Our Planet*. Cambridge, MA: Harvard University Press, 2007.

Osborne, Roger, Donald Tarling, and Stephen J. Gould, eds. *The Historical Atlas of the Earth: A Visual Celebration of Earth's Physical Past*. New York: Henry Holt, 1996.

Ostfeld, Richard W. *Lyme Disease: The Ecology of a Complex System*. New York: Oxford University Press, 2010.

Palmer, Douglas. *Origins: Human Evolution Revealed*. London: Mitchell Beazley, 2010.

Preston, Richard. *The Hot Zone: A Terrifying True Story*. New York: Random House, 1994.

Ridley, Matt. *Genome: The Autobiography of a Species in 23 Chapters*. New York: HarperCollins, 2000.

Roberts, Callum. *The Unnatural History of the Sea*. Washington, DC: Island Press/Shearwater Books, 2007.

Schlesinger, William, and Emily Bernhardt. *Biogeochemistry: An Analysis of Global Change*. San Diego: Academic Press, 2013.

Stager, Curt. *Deep Future: The Next 100,000 Years of Life on Earth*. New York: Thomas Dunne Books/St. Martin's Press, 2011.

Steinbeck, John. *The Log from the Sea of Cortez*. New York: Viking Press, 1951.

Sues, Hans-Dieter, and Nicholas C. Fraser. *Triassic Life on Land: The Great Transition*. New York: Columbia University Press, 2010.

Tattersall, Ian. *Masters of the Planet: The Search for Our Human Origins*. New York: Palgrave Macmillan, 2012.

Terborgh, John, and James A. Estes. *Trophic Cascades: Predators, Prey, and the Changing Dynamics of Nature*. Washington, DC: Island Press, 2010.

Thornton, Ian. *Krakatau: The Destruction and Reassembly of an Island Ecosystem*. Cambridge, MA: Harvard University Press, 1996.

Wallace, William McDonald. *Techno-Cultural Evolution: Cycles of Creation and Conflict*. Washington, DC: Potomac Books, 2006.

Ward, Peter. *Future Evolution: The Illuminated History of Life to Come*. New York: W. H. Freeman, 2001.

Ward, Peter, and Donald Brownlee. *The Life and Death of Planet Earth: How the Science of Astrobiology Charts the Ultimate Fate of Our World*. New York: Times Books, 2002.

Weiner, Jonathan. *The Beak of the Finch: A Story of Evolution in Our Time*. New York: Alfred A. Knopf, 1994. Winner of the Pulitzer Prize.

Weisman, Alan. *The World Without Us*. New York: Thomas Dunne Books/St. Martin's Press, 2007.

Wilson, Edward O. *The Future of Life*. New York: Alfred A. Knopf, 2002.

Winchester, Simon. *Krakatoa: The Day the World Exploded, August 27, 1883*. New York: HarperCollins, 2003.

Zalasiewicz, Jan. *The Earth After Us: What Legacy Will Humans Leave in the Rocks?* Oxford, UK: Oxford University Press, 2008.

Zubrin, Robert. *The Case for Mars: The Plan to Settle the Red Planet and Why We Must*. New York: Free Press, 1996.

ARTICLES

Acuña, José Luis, Ángel López-Urrutia, and Sean Colin. "Faking Giants: The Evolution of High Prey Clearance Rates in Jellyfish." *Science* 333 (September 16, 2011).

Albright, Rebecca, Benjamin Mason, Margaret Miller, and Chris Langdon. "Ocean acidification compromises recruitment of the threatened caribbean coral *Acropora palmata*." *Proceedings of the National Academy of Sciences* 107, no. 47 (November 23, 2010): 20400–404.

Alexander, Vera, Patricia Miloslavich, and Kristin Yarinchik. "The Census of Marine Life—Evolution of Worldwide Marine Biodiversity Research." *Marine Biodiversity* 41, no. 4 (March 1, 2011): 545–54.

Allen, J. P., M. Nelson, and A. Alling. "The Legacy of Biosphere 2 for the study of biospherics and closed ecological systems." *Advances in Space Research* 31, no. 7 (2003): 1629–40.

Amos, Jonathan. "Dirty Stars Hint at Sun's Future." BBC News, May 9, 2013.

Anderson, Ross. "We're Underestimating the Risk of Human Extinction." *The Atlantic*, March 2012, http://www.theatlantic.com/technology/archive/2012/03/were-underestimating-the-risk-of-human-extinction/253821/.

Archer, Steve, David S. Schimel, and Elizabeth A. Holland. "Mechanisms of shrubland expansion: land use, climate or CO_2?" *Climatic Change* 29 (1995): 91–99.

Arroyo-Kalin, Manuel. "Slash-burn-and-churn: landscape history and crop cultivation in Pre-Columbian Amazonia." *Quaternary International* 249 (February 6, 2012): 4–18.

Bardin, Jon. "Beneath 50-foot ice layer, an Antarctic lake full of life." *Los Angeles Times*, November 27, 2012.

Barnosky, A. D. "Megafauna biomass tradeoff as a driver of quaternary and future extinctions." *Proceedings of the National Academy of Sciences* 105 (August 12, 2008): 11543–48.

Barnosky, Anthony D., Nicholas Matzke, Susumu Tomiya, Guinevere O. U. Wogan, Brian Swartz, Tiago B. Quental, Charles Marshall, et al. "Has the Earth's Sixth Mass Extinction Already Arrived?" *Nature* 471 (March 3, 2011): 51–57.

Baron, Roy C., Joseph B. McCormick, and Osman A. Zubeir. "Ebola Virus Disease in Southern Sudan: Hospital Dissemination and Intrafamilial Spread." *Bulletin of the World Health Organization* 61, no. 6 (1985): 997–1003.

Benanav, Michael. "Through the Eyes of the Maasai." *New York Times*, August 9, 2013.

Bennett, Liz. "For Sale: Black Market Orangutan Babies." Wildlife Conservation Society, March 15, 2012, http://e.wcs.org/site/MessageViewer?em_id=20162.0&dlv_id=25702.

Bergin, Chris. "Fobos-Grunt Ends Its Misery via Re-entry." NASA, January 15, 2012, http://www.nasaspaceflight.com/2012/01/fobus-grunt-ends-its-misery-via-re-entry/.

Biello, David. "The Origin of Oxygen in Earth's Atmosphere." *Scientific American*, August 19, 2009.

Blackburn, Terrence J. "Zircon U-Pb Geochronology Links End-Triassic Extinction with the Central Atlantic Magmatic Province." *Science* 340, no. 6135 (May 24, 2013): 941–45.

Bostrom, Nick. "The Future of Human Evolution." Published in *Death and Anti-Death: Two Hundred Years After Kant, Fifty Years After Turing*, ed. Charles Tandy (Ria University Press: Palo Alto, CA, 2004), 339–71.

Brunner, Jesse, et al. "An experimental test of competition among mice, chipmunks, and squirrels in deciduous forest fragments." *PLOS ONE*, June 18, 2013.

Bryson, Donna. "Wildlife Drive Grows from Superfund Site." *San Francisco Chronicle*, October 19, 2012. www.sfgate.com.

"Burgess Shale: Strange Creatures—A Burgess Shale Fossil Sampler." Smithsonian National Museum of Natural History, http://paleobiology.si.edu/burgess. This is a good source for Burgess Shale animals including *Opabinia*, *Amiskwia*, and *Anomalocaris*.

Burgess, Seth D., Samuel Bowring, and Shu-zhong Shen. "High-precision timeline for Earth's most severe extinction." *Proceedings of the National Academy of Sciences*, February 10, 2014.

Bush, Mark B., Miles R. Silman, Dunia H. Urrego. "48,000 Years of Climate and Forest Change in a Biodiversity Hot Spot." *Science* 303, no. 5659, (February 6, 2004): 827–29.

Cardinale, Bradley J. "Biodiversity loss and its impact on humanity." *Nature* 486, (June 7, 2012): 941–45.

Carr, Steve. "UNM Researchers Explore Evolution of World's Mammals Over the Past 100 Million Years." University of New Mexico, November 25, 2010.

Chan, F. "Emergence of Anoxia in the California Current Large Marine Ecosystems." *Science* 319, February 15, 2008, sciencemag.org.

Cisneros-Montemayor, Andres, et al. "Global economic value of shark ecotourism: implications for conservation." *Oryx*, July 2013.

Cohen, Chad. "Bioinvasion: From Old World to New." *National Geographic News*, January 23, 2001.

Conde, Dalia A. "Modeling male and female habitat difference for jaguar conservation." *Biological Conservation*, May 31, 2010.

Coughlan, Sean. "Are humans going to become extinct?" BBC News, April 23, 2013.

Cross, Wyatt F., et al. "Food-web dynamics in a large river discontinuum." *Ecological Monographs* 83, no. 3 (August 2013).

Dalhousie University. "Shark Fisheries Globally Unsustainable: New Study—Researchers Estimate 100 Million Sharks Die Every Year." *Newswise*, Dalhousie University, March 1, 2013.

Dell'Amore, Christine. "Next Ice Age Delayed by Global Warming, Study Says." *National Geographic News*, September 3, 2009.

Deneen, Sally. "Feds Slash Colorado River Release to Historic Lows." *National Geographic*, August 16, 2013.

Deng, Tao, et al. "Out of Tibet: Pliocene Wooly Rhino Suggests High-Plateau Origin of Ice Age Megaherbivores." *Science* 333, no. 6047 (September 2, 2011): 1285–88.

Donlan, Josh C., et al. "Pleistocene Rewilding: An Optimistic Vision for 21st Century Conservation." *The American Naturalist*, November 2006.

"Doomsday Clock Moves One Minute Closer to Midnight. It is Now 5 Minutes to Midnight." *Bulletin of the Atomic Scientists*, January 10, 2012.

Dreifus, Claudia. "Studying Evolution with an Eye on the Future." *New York Times*, July 30, 2012.

Dupuis II, Alan P. "Isolation of deer tick virus (Powassan virus, lineage II) from Ixodes scapularis and detection of antibody in vertebrate hosts sampled in the Hudson Valley, New York State." *Parasites and Vectors*, July 15, 2013.

Ehrlich, Paul and Anne Ehrlich. "The Population Bomb Revisited." *The Electronic Journal of Sustainable Development* 1, no. 3 (2009). Population-Bomb-Revisited-Paul-Ehrlich_20096-6.pdf, www.ejsd.org.

Esters, James A., et al. "Using Ecological Function to Develop Recovery Criteria for Depleted Species: Sea Otters and Kelp." *Conservation Biology* (2010).

Estes, James A., et al. "Causes and consequences of marine mammal population declines in southwest Alaska: a food-web perspective." *Phil. Trans. R. Soc.* (June 25, 2009).

Estes, James A., et al. "Trophic Downgrading of Planet Earth." *Science* 333, no. 6040 (July 15, 2011): 301–6.

Estrada-Pena, et al. "Effects of environmental change on zoonotic disease risk: an ecological primer." *Trends in Parasitology* 30, no. 4 (April 2014), 205–14.

European Geoscience Union. "The oldest ice core. Finding a 1.5 million-year record of Earth's climate." *EurekAlert!* November 5, 2013.

European Space Agency. "Choosing the right people to go to Mars." January 13, 2013. http://www.esa.int/Our_Activities/Human_Spaceflight/Mars500/Choosing_the_right_people_to_go_to_Ma.

Evans, Alistair R., et al. "The maximum rate of mammal evolution." *Proceedings of the National Academy of Sciences* (October 1, 2011), http://www.pnas.org/content/early/2012/01/26/1120774109.abstract.

Evolution UC Berkeley. "Ensatina's basic story was laid out by Robert Stebbins 30 years before Tom was born in 1977." Evolution Web Site, http://evolution.berkeley.edu.

"Experts Reaffirm Asteroid Impact Caused Mass Extinction," University of Texas Austin. March 4, 2010, http://www.utexas.edu/news/2010/03/04/mass_extinction/.

Fecht, Sarah. "Going Wireless and Restoring Memories: The Incredible Future of Brain Implants." *Popular Mechanics*, June 17, 2013.

Fernández-Busquets, Xavier, et al. "Self-recognition and Ca^{2+}-dependent carbohydrate–carbohydrate cell adhesion provide clues to the Cambrian Explosion." *Molecular Biology and Evolution* 2b, no. 11 (2009): 2551.

Fienberg, Rick. "JPL: 'Mount Sharp' on Mars Links Geology's Past & Future." *American Astronomical Society*, March 28, 2012.

———. "JPL: Mars Rover Carries Device for Underground Scouting." *American Astronomical Society*, October 20, 2011.

———. "MIT: Are you a Martian? We could all be, scientists say." *American Astronomical Society*, March 23, 2011.

Flynn, Sian. "The Race to the South Pole." *BBC History in Depth*. March 3, 2011.

Frenzen, P. "30 Years Later, Forest Rebirth is Well Under Way." US Forest Service, Mt. Saint Helens Volcanic Monument. http://www.fs.usda.gov/main/mountst helens/learning/nature-science.

———. "How is plant recovery likely to proceed in future?" (Mt. Saint Helens) USDA Forest Service, 1994.

———. "Life Returns: Frequently Asked Questions about Plant and Animal Recovery Following the 1980 Eruption." US Forest Service, Mt. Saint Helens Volcanic Monument. http://www.fs.usda.gov/mountsthelens.

Gaffin, Stuart R., Cynthia Rosenzweig, and Angela Y. Y. Kong. "Adapting to climate change through urban green infrastructure." *Nature Climate Change* 2 (October 2012).

Gallessich, Gail. "Major Study of Ocean Acidification Helps Scientist Evaluate Effects of Atmospheric Carbon Dioxide on Marine Life." Public Affairs & Communications, University of California, Santa Barbara, January 23, 2013, http://www .ia.ucsb.edu/pa/display.aspx?pkey=2618.

Gamo, Toshitaka, Harue Masuda, Toshiro Yamanaka, Kei Okamura, Junichiro Ishibashi, Eichiro Nakayama, Hajime Obata, et al. "Discovery of a new hydrothermal venting site in the southernmost Mariana arc." *Geochemical Journal* 38 (2004): 527–34.

Garrabou, J., R. Coma, N. Bensoussan, M. Bally, P. Chevaldonné, M. Cigliano, D. Diaz, et al. "Mass Mortality in Northwestern Mediterranean Rocky Benthic Communities: Effects of the 2003 Heat Wave." *Global Change Biology* 15, no. 5 (May 2009): 1090–1103.

Garrity, Lyn. "Evolution World Tour: Mount St. Helens Washington: Over thirty years after the volcanic eruptions, plant and animal life has returned to the disaster site, a veritable living laboratory." *Smithsonian*, January 2012.

Garthwaite, Josie. "Into the Permian: In Mongolia, 298-million-year-old plant fossils paint a portrait of what life was like long before the dinosaurs." *Discover*, November 2012.

Gettleman, Jeffrey. "Elephants Dying in Epic Frenzy as Ivory Fuels Wars and Profits." *New York Times*, September 3, 2012.

Gillis, Justin. "Deep Thinking About the Future of Food." *New York Times*, October 12, 2011.

Gilly, William F., and Unai Markaida. "Perspectives on *Dosidicus gigas* in a changing world." Workshop: The role of squid in open ocean ecosystems, Hawaii, November 16–17, 2006.

Gilly, William. "Searching for the Spirits of the Sea of Cortez." *Steinbeck Studies* 15, no. 2 (Fall 2004).

Gorenflo, L. J., Suzanne Romaine, Russell A. Mittermeier, and Kristen Walker-Painemilla. "Co-occurrence of linguistic and biological diversity hotspots and high biodiversity wilderness areas." *Proceedings of the National Academy of Sciences* 109, no. 21 (May 7, 2012): 8032–57.

Grieco, Theresa M., et al. "A Modular Framework Characterizes Micro- and Macroevolution of Old World Monkey Dentitions." *Evolution*, January 2013.

Gugliotta, Guy. "The Great Human Migration: Why humans left their African homeland 80,000 years ago to colonize the world." *Smithsonian*, July 2008.

Hall, Stephen S. "The Other Humans: Neanderthals Revealed." *National Geographic*, October 2008.

Hammer, Joshua. "The Hunt for Ebola." *Smithsonian*, November 2012.

Handwerk, Brian. "Do Hammerheads Follow Magnetic Highways to Migration?" *National Geographic News*, June 6, 2002.

Harris, John M. "Bones from the Tar Pits." *Natural History*, June 2007.

Hawks, John, et al. "Recent acceleration of human adaptive evolution." *Proceedings of the National Academy of Sciences* (December 26, 2007).

Hester, Keith, et al. "Unanticipated consequences of ocean acidification: A noisier ocean at lower pH." *Geophysical Research Letters* 35 (October 1, 2008).

Hof, Robert D. "Deep Learning: With massive amounts of computational power, machines can now recognize objects and translate speech in real time. Artificial intelligence is finally getting smart." *MIT Technology Review*, April 23, 2013.

Hoffman, Hillel J. "The Permian Extinction." *National Geographic*, September 2000.

Hollingham, Richard. "Building a new society in space." BBC Future, March 18, 2013.

Holloway, Marguerite. "Cuttlefish Say It With Skin." *Natural History*, April 2000.

"How to Get Along for 500 Days Alone Together." *BBC News Magazine*, March 1, 2013, http://www.bbc.com/news/magazine-21619765.

"How Volcanoes Work: Krakatau, Indonesia," Geology, San Diego State University, http://www.geology.sdsu.edu/how_volcanoes_work/Krakatau.html.

Institut de Recherche pour le Développement (IRD). "Boom in jellyfish: Overfishing called into question." *ScienceDaily*, May 3, 2013.

Intergovernmental Science-policy Platform on Biodiversity and Ecosystem Services (IPBES), "Even farm animal diversity is declining as accelerating species loss threatens humanity." *EurekAlert!* May 27, 2013.

Irmis, Randall B., Sterling J. Nesbitt, and Hans-Dieter Sues. "Early Crocodylomorpha." From "Anatomy, Phylogeny and Palaeobiology of Early Archosaurs and their Kin." *Geological Society of London Special Publications* 379: 275–302. First published online June 11, 2013, http://dx.doi.org/10.1144/SP379.24.

Jackson, Robert B., et al. "Natural gas pipeline leaks across Washington, DC." *Environmental Science & Technology* 48, no. 3 (January 16, 2014): 2051–58.

James, Ian. "Officials discuss solutions for sea: Officials, residents optimistic concerted efforts can succeed." *Desert Sun*, April 27, 2013.

John F. Kennedy Presidential Library and Museum. "Nuclear Test Ban Treaty." http://www.jfklibrary.org/JFK/JFK-in-History/Nuclear-Test-Ban-Treaty.aspx.

Johnson, Donald: "Origins of Modern Humans: Multiregional or Out of Africa?" *Action Bioscience*, May 2001.

Jubinsky, G. "Chinese Tallow (*Sapium sebiferum*)." Florida Department of Environmental Protection, Bureau of Aquatic Plant Management. Pub. No. TSS-93-03, 1995.

Kaufman, Frederick. "The Second Green Revolution: An Alliance of Organic Farmers and Genetic Engineers." *Popular Science*, posted January 20, 2011.

Kaufman, Leslie. "Zoos' Bitter Choice: To Save Some Species, Letting Others Die." *New York Times*, May 27, 2012.

Keesing, Felicia, and Truman P. Young. "Cascading Consequences of the Loss of Large Mammals in an African Savanna." *BioScience*, May 7, 2014.

Khan, Amina. "Study: Mars could have held watery underground oases for life." *Los Angeles Times*, January 21, 2013.

Kluger, Jeffrey. "Live from Mars: A one-ton Rover can teach us a lot about the red planet—and the blue one too." *Time*, August 30, 2012.

Knoll, A. H., R. K. Bambach, D. E. Canfield, and J. P. Grotzinger. "Comparative Earth History and Late Permian Mass Extinction." *Science* 273, no. 5274 (July 26, 1996): 452–57.

Knoll, Andrew H., Richard K. Bambach, Jonathan L. Payne, Sara Pruss, and Woodward W. Fischer. "Paleophysiology and End-Permian Mass Extinction." *Earth and Planetary Science Letters* 256, nos. 3–4 (April 30, 2007): 295–313, www.sciencedirect.com.

Kolbert, Elizabeth. "Sleeping with the Enemy: What Happened Between the Neanderthals and Us?" *New Yorker*, August 15 and 22, 2011.

Kotler, Steven. "Evolution's Next Stage: Driven by technological advances, humans are changing faster than ever. Coming soon: our next stage, Homo evolutus." *Discover*, March 2013.

Krajick, Kevin. "Seeking the Deadly Roots of the Dinosaurs' Ascent." *State of the Planet: Blogs from Earth Institute*, August 17, 2012.

"Krill face deadly cost of ocean acidification," Australian Antarctic Division, October 13, 2010, www.anarctica.gov.au.

Kroeker, Kristy J., Fiorenza Micheli, Maria Cristina Gambi, and Todd R. Martz. "Divergent Ecosystem Responses Within a Benthic Marine Community to Ocean Acidification." *Proceedings of the National Academy of Sciences* 108, no. 35 (August 30, 2011): 14515–20.

Kumar, Alexander. "Viewpoint: When will we send humans to Mars?" *BBC News: Science and Environment*, September 21, 2012.

Kunzig, Robert. "Seven Billion: Special Series." *National Geographic*, January 2011.

Lamont-Doherty Earth Observatory. "Megavolcanoes Tied to Pre-Dinosaur Extinction: An Apparent Sudden Climate Shift Could Have Analog today." Columbia University, Earth Institute, March 20, 2013.

Langford, Kate. "400,000 farmers in Africa use fertilizer trees to improve food security." World Agroforestry Centre, October 14, 2011, http://www.worldagroforestrycentre.org/newsroom/highlights/400000-farmers-africa-use-fertilizer-trees-improve-food-security.

Lawrence, Robert S. "The Rise of Antibiotic Resistance: Consequences of FDA's Inaction." *The Atlantic*, January 23, 2012.

Le Page, Michael, "Reshaping Eden: the future of biodiversity." *New Scientist*, April 24, 2010.

Lemonick, Michael D. "Super-Crocodiles May Have Dined on Dinosaurs." *Time*, November 23, 2009.

Lewis, Richard. "Land animals, ecosystems walloped after Permian dieoff." Brown University, October 25, 2011.

Lidgard, Scott, Peter J. Wagner, and Matthew A. Kosnik. "The Search for Evidence of Mass Extinction." *Natural History*, September 2009.

Lordkipanidze, David, et al. "A Complete Skull from Dmanisi, Georgia, and the Evolutionary Biology of Early *Homo*." *Science* 342, no. 6156 (October 18, 2013): 326–31.
Lucas, Tim. "Looking Back, and Forward, at Changes in Soil Science." *Duke Environment*, Duke University, May 21, 2012.
Maestre, Fernando, et al. "Plant Species Richness and Ecosystem Multifunctionality in Global Drylands." *Science* 335, no. 6065 (January 13, 2012): 214–18.
Mangan, Scott A., et al. "Negative plant–soil feedback predicts tree-species relative abundance in a tropical forest." *Nature*, June 25, 2010, http://www.nature.com/nature/journal/v466/n7307/full/nature09273.html.
Mann, Charles C. "State of the Species: Does success spell doom for *Homo sapiens*?" *Orion Magazine*, November/December 2012.
Marean, Curtis W. "When the Sea Saved Humanity." *Scientific American*, August 2010.
Markoff, John. "Obama Seeking to Boost Study of Human Brain." *New York Times*, February 17, 2013.
Marsh, Bill. "Are We in the Midst of a Sixth Mass Extinction? A Tally of Life Under Threat." *New York Times*, Sunday Review, Opinion Pages, June 1, 2012.
Marshall, Michael. "Animals are already dissolving in Southern Ocean." *New Scientist*, November 25, 2012.
Martin, Ronald, and Antonietta Quigg. "Tiny Plants That Once Ruled the Seas: Why did animal life in the ocean explode with diversity about 250 million years ago? The driving force may have been the rise of phytoplankton." *Scientific American*, June 2013.
Martin, William and Michael J. Russell. "On the origin of biochemistry at an alkaline hydrothermal vent." *Philosophical Transactions of the Royal Society of London, Series B: Biological Sciences* 362, no. 1486 (October 29, 2007): 1887–925.
McGill University. "Biodiversity critical for maintaining multiple 'ecosystem services.'" *ScienceDaily*, August 19, 2011.
Meachen, Julie A., and Joshua X. Samuels. "Evolution in Coyotes (*Canis latrans*) in Response to the Megafaunal Extinctions." *Proceedings of the National Academy of Sciences* 109, no. 11 (March 13, 2012): 4191–96.
Mitchell, Alanna. "Life in the Sea Found its Fate in a Paroxysm of Extinction." *New York Times*, April 30, 2012.
Mohan, Geoffrey. "Volcano-induced die-off paved way for dinosaurs, study suggests." *Los Angeles Times*, March 21, 2013.
Moreau, Claudia, et al. "Deep human genealogies reveal a selective advantage to be on an expanding wave front." *Science* 334, no. 6059 (November 25, 2011): 1030.
Morelle, Rebecca. "Malaria 'spreading to new altitudes': Warmer temperatures are causing malaria to spread to higher altitudes, a study suggests." *BBC News Health*, March 6, 2014.
Museum of New Zealand. "Colossal Squid: Hooks and suckers." Te Papa Museum, April 30, 2008, http://squid.tepapa.govt.nz/anatomy/article/the-arms-and-tentacles.
Myers, Norman and Andrew H. Knoll. "The biotic crisis and the future of evolution." *Proceedings of the National Academy of Sciences*, May 8, 2001.
Myers, Norman, et al. "Biodiversity hotspots for conservation priorities." *Nature* 403 (February 24, 2000).
NASA/Jet Propulsion Laboratory. "Is a sleeping climate giant stirring in the Arctic?" *ScienceDaily*, June 11, 2013.
National Park Service. "Geology fieldnotes, Guadalupe Mountains National Park, Texas." http://www.nature.nps.gov/geology/parks/gumo.

National Science Foundation. "Mammals grew 1,000 times larger after the demise of the dinosaurs." NSF, November 25, 2010.

Natural History Museum of Los Angeles County. "Woolly rhino fossil discovery in Tibet provides important clues to evolution of Ice Age giants." *ScienceDaily*, September 2, 2011.

Nature Conservancy. "Palmyra: A Marine Wilderness: Palmyra is that rare place where top predators such as sharks still dominate the reef ecosystem." www.nature.org.

Netburn, Deborah. "Did life on Earth start on Mars? A scientist lays out the evidence." *Los Angeles Times*, August 29, 2013.

"New Research Rejects 80-Year Theory of 'Primordial Soup' as the Origin of Life." *ScienceDaily*, February 3, 2010.

Njau, Jackson K., and Robert Blumenschine. "A diagnosis of crocodile feeding traces on larger mammal bone, with fossil examples from Plio-Pleistocene Olduvai Basin, Tanzania." *Journal of Human Evolution* 50 (2006): 142–62. Available online at www.sciencedirect.com.

———. "Crocodylian and mammalian carnivore feeding traces on hominid fossils from FLK 22 and FLK NN 3, Plio-Pleistocene Olduvai Gorge, Tanzania." *Journal of Human Evolution* 63, no. 2 (August 2012): 408–17. Available online at www.sciencedirect.com.

NSBRI National Space Biomedical Research Institute. "How the Human Body Changes in Space." http://www.nsbri.org/DISCOVERIES-FOR-SPACE-and-EARTH/The-Body-in-Space/.

Nuwer, Rachel. "Fukushima vs. Chernobyl: How Have Animals Fared?" *New York Times*, July 12, 2012.

O'Connor, Anahad. "No Fins? No Problem: Jellyfish Have Their Ways." *New York Times*, September 30, 2011.

O'Hara, Carolyn. "America's energy future: The search for alternatives to oil and gas." *The Week*, September 6, 2013.

Ostfeld, Richard. "Human health hinges on species diversity." *PoughkeepsieJournal.com*, August 19, 2007.

Ostfeld, Richard S. and Robert D. Holt. "Are predators good for your health? Evaluating evidence for top-down regulation of zoonotic disease reservoirs." *Frontiers in Ecology and the Environment* 2, no. 1 (2004): 13–20.

"Oxygen-Free Early Oceans Likely Delayed Rise of Life on Planet." University of California, Riverside, January 10, 2011, http://newsroom.ucr.edu/2520.

Overbye, Dennis. "2 Good Places to Live, 1,200 Light-Years Away." *New York Times*, April 18, 2013.

Pacific Northwest Research Station. "Mount St. Helens 30 Years Later: A Landscape Reconfigured." http://www.fs.fed.us/pnw/mtsthelens/.

Pardi, Melissa I., and Felisa A. Smith. "Paleoecology in an Era of Climate Change: How the Past Can Provide Insights into the Future." In Louys, Julien, ed. *Paleontology in Ecology and Conservation*. Springer, 2012.

Pastino, Blake de. "Giant Jellyfish Invade Japan." *National Geographic News*, October 28, 2010.

Payne, Jonathan L. "The evolutionary consequences of oxygenic photosynthesis: a body size perspective." *Photosynth Res*, September 7, 2010.

Pearson, Richard. "Protecting Many Species to Help Our Own." *New York Times*, June 1, 2012.

Phillips, Nathan, et al. "Mapping urban methane pipeline leaks: Methane leaks across Boston." *Environmental Pollution* 173 (2013): 1–4.

Quammen, David. "Anticipating the Next Pandemic." *New York Times*, September 22, 2012.

Quillen, Lori. "Dams Destabilize River Food Webs: Lessons from the Grand Canyon." Cary Institute of Ecosystem Studies, August 20, 2013.

———. "Black-legged Ticks Linked to Encephalitis in New York State." Cary Institute of Ecosystem Studies, July 15, 2013, www.caryinstitute.org.

Raffaele, Paul. "Speaking Bonobo: Bonobos Have an Impressive Vocabulary, Especially When It Comes to Snacks." *Smithsonian*, November 2006, http://www.smithsonianmag.com/science-nature/speaking-bonobo-134931541/.

"Rearing cattle produces more greenhouse gases than driving cars, UN report warns." UN News Centre, November 29, 2006, http://www.un.org/apps/news/story.asp?newsID=20772.

Regalado, Antonio. "Genome Hunters Go After Martian DNA." *Technology Review*, October 18, 2012, http://www.technologyreview.com/news/429662/genome-hunters-go-after-martian-dna/.

Revkin, Andrew. "Papers Find Mixed Impacts on Ocean Species from Rising CO_2." *New York Times*, August 26, 2013.

Rich, Nathaniel. "Can Jellyfish Unlock the Secret of Immortality?" *New York Times*, November 28, 2012.

Richter, Daniel and Dan H. Yaalon. "The changing model of soil, revisited." *Soil Science Society of America Journal* 76, no. 3 (May/June 2012).

Richter, Daniel, et al. "Evolution of Soil and Ecosystem Research at the Calhoun Experimental Forest." *Research for the Long Term*. Springer, 2013.

Ripple, W. J. and Blaire Van Valkenburgh. "Liking top-down forces to the Megafaunal extinctions." *Bioscience* 60 (2010): 516–26.

Roberts, Mitchell. "New 'SARS-like' coronavirus identified by UK officials." *BBC News/Health*, September 24, 2012.

Roemer Gary W., Matthew W. Gompper, and Blaire Van Valkenburgh. "The Ecological Role of Mammalian Mesocarnivore." *BioScience*, February 2009.

Rosi-Marshall, Emma. "Colorado River can be revived." *Poughkeepsie Journal*, September 11, 2011.

Rothamsted Research. "Rothamsted Research: where knowledge grows." *Science Strategy*, 2012 to 2017. Downloaded from www.rothamsted.ac.uk.

Rothamsted Research. "Using the world's oldest field experiment to detect nuclear fallout and other environmental changes." Downloaded from www.rothamsted.ac.uk.

Roubinette, Tom. "The mammoth's lament: UC Research shows how cosmic impact sparked devastating climate change." *EurekAlert!* May 20, 2013.

Sabin, Paul. "Betting on the Apocalypse." *New York Times*, September 7, 2013.

Sagarin, Raphael D. "Remembering the Gulf: Changes in the marine communities of the Sea of Cortez since the Steinbeck and Ricketts expedition of 1940." *Frontiers in Ecology* 6, no. 7 (2008): 372–79.

"Short History of Los Angeles, A." Communication Studies, University of California, Los Angeles, http://cogweb.ucla.edu/Chumash/LosAngeles.html.

Smith, Felisa A., Scott M. Elliot, and Kathleen Lyons. "Methane emissions for extinct megafauna." *Nature Geoscience*, May 23, 2010.

Smith, J. T., N. J. Willey, and J. T. Hancock. "Low-dose ionizing radiation produces too few reactive oxygen species to directly affect antioxidant concentrations in cells." *Biology Letters*, 2012.

Smithsonian, Department of Paleontology. "The Triassic, Overview; Extinction and Recovery." National Museum of Natural History, http://paleobiology.si.edu.

Stebbins, R. C. "Speciation in salamanders of the plethodontid genus *Ensatina*." *University of California Publications in Zoology* 48 (1949): 377–526.
Steel, Bill. "Before cells, biochemicals may have combined in clay." *Cornell Chronicle*, November 7, 2013, www.news.cornell.edu.
Steudel, Bastian, et al. "Biodiversity effects on ecosystem functioning change along environmental stress gradients." *Ecology Letters*, December 2012.
Stewart, Julia S. "Onshore-offshore movement of jumbo squid (*Dosidicus gigas*) on the continental shelf." *Deep-Sea Research II*, May 24, 2014.
Stramma, L. A. Oschlies and S. Schmidtko. "Mismatch between observed and modeled trends in dissolved upper-ocean oxygen over the last 50 yr." *Biogeosciences* 9 (2012): 4045–57.
Stramma, Lothar. "Expanding Oxygen-Minimum Zones in the Tropical Oceans." *Science*, May 2, 2008.
———. "Ocean oxygen minimum expansion and their biological impacts." *Deep Sea Research Part I: Oceanographic Research Papers*, April 2010, 587–95.
Suda-King, Chikako. "Do orangutans (*Pongo pygmaeus)* know when they do not remember?" *Animal Cognition* 11 (2008): 21–42.
Switek, Brian. "Broken teeth tell of tough times for Smilodon." *ScienceBlogs*, February 15, 2010.
Tangley, Laura. "New Mammals in Town. There's plump, furry, beaked, and some of them bark." *US News & World Report*, June 2, 1997.
Ted Talks. "Will our kids be a different species?" Juan Enriquez, April 2012.
Tennesen, Michael. "Avatar Acts: When the Matrix has you, what laws apply to settle conflicts?" *Scientific American*, July 2009.
———. "Mars: Remembrance of Life Past. The Viking probes failed to find living organisms on Mars, but new studies suggest that the Red Planet may not have always been dead." *Discover*, July 1989.
———. "Python Predation: Big snakes poised to change US ecosystems." *Scientific American*, January 20, 2010.
———. "Stars in their Eyes: The exquisite telescopes crafted by Alvan Clark and his sons helped make the last half of the 19th century a golden age of astronomy." *Smithsonian*, October 2001.
———. "Black Gold of the Amazon: Fertile, charred soil created by pre-Columbian peoples sustained late settlements in the rain forest. Secrets of that ancient 'dark earth' could help solve the Amazon's ecological problems today." *Discover Magazine*, April 2007.
———. "Can the Military Clean Up Its Act? The military is working at becoming friend rather than foe to the wildlife on its lands—but toxic hot spots complicate the mission." *National Wildlife*, October 1, 1993.
———. "Deep Sea Divers: How low can marine animals go?" *Wildlife Conservation*, June 2005.
———. "Expedition to the Clouds." *International Wildlife Magazine*, March/April 1998.
———. "Humboldt Squid: Masters of Their Universe." *Wildlife Conservation*, February 2009.
———. "Mountains Under the Sea." *National Wildlife*, September/October 2000.
———. "Myth of the Monster: Deadly? Of course, but studies confirm the great while shark is a remarkable creature that poses only a rare threat to people." *National Wildlife*, October/November 1989.
———. "Outsmarting the Competition: When it comes to intelligence and personality, the giant Pacific octopus shines." *National Wildlife*, December 2002.

———. "Phosphorus Fields: Phosphorus and nitrogen fertilizers drive modern agriculture, but they are also poisoning the planet." *Discover*, December 2009.

———. "Testing the Depth of Life: Northern elephant seals migrate farther than any other mammal, spending much of their time at bone-crushing depths. How do they do it?" *National Wildlife*, February/March 1999.

———. "The White Shark Cafe." *National Wildlife*, August/September 2011.

———. "The Strange Forests that Drink—and Eat—Fog." *Discover*, April 2009.

———. "Tuning in to Humpback Whales." *National Wildlife*, February/March 2002.

———. "Turning to Dust: Around the globe, grasslands are turning to desert and free-flowing bits of dirt and rock are remaking the environment." *Discover*, May 2010.

———. "Uphill Battle." *Smithsonian*, August 2006.

———. "Waiting to Inhale: Deep-Ocean Low-Oxygen Zones Spreading to Shallower Coastal Waters." *Scientific American*, February 23, 2010.

———. "When Juniper and Woody Plants Invade, Water May Retreat." *Science* 322 (December 12, 2008).

Terborgh, J., et al. "Ecological Meltdown in Predator-Free Forest Fragments." *Science*, 2001.

Than, Ker. "Drug-filled Mice Airdropped Over Guam to Kill Snakes." *National Geographic News*, September 24, 2010.

The College of Physicians of Philadelphia. "The History of Vaccines: Yellow Fever." 2014, www.historyofvacciens.org.

The National Museum; Royal Navy (UK). "Biography: Captain Robert Scott." *Royal Naval Museum Library*, 2004.

Thornton, Ian. "Figs, frugivores and falcons: An aspect of the assembly of mixed tropical forest on the emergent volcanic island, Anak Krakatau." *South Australian Geographical Journal* 93 (1994): 3–21.

Turner, Pamela S. "Darwin's Jellyfishes." *National Wildlife*, August/September 2006.

Understanding Evolution, UC Berkeley. "Biogeography: Wallace and Wegener." http://evolution.berkeley.edu/evolibrary/article/history_16.

———. "How did life originate?" http://evolution.berkeley.edu/evolibrary/.

UNESCO, Culture, World Heritage Centre. "Ngorongoro Conservation Area: Outstanding Universal Value." http://whc.unesco.org/en/list/39.

University of Arizona. "UA Science Biosphere 2." http://b2science.org/.

University of British Columbia. "Sharks worth more in the ocean than on the menu." *ScienceDaily*, May 30, 2013.

University of Massachusetts at Amherst. "Ice-free Arctic may be in our future, say UMass-Amherst, international researchers." *EurekAlert!* May 9, 2013.

University of Portsmouth. "Wildlife thriving after nuclear disaster? Radiation from Chernobyl and Fukushima nuclear accidents not as harmful to wildlife as feared." *ScienceDaily*, April 11, 2012.

University of Tennessee. "Crocodilians can climb trees and bask in tree crowns." *ScienceDaily*, February 10, 2014.

University of Utah. "Are Humans Evolving Faster? Findings Suggest We Are Becoming More Different, Not Alike." University of Utah Public Relations, December 10, 2007.

University of Waterloo. "Dramatic thinning of Arctic lake ice cuts winter ice season by 24 days compared to 1950." *ScienceDaily*, February 3, 2014.

Valkenburgh, Blaire Van, et al. "Cope's Rule, Hypercarnivory, and Extinction in North American Canids." *Science*, October 1, 2004.

Van Valkenburgh, Blaire. "Tough Times in the Tar Pits." *Natural History*, April 1994.
Vetter, Russ. "Predatory interactions and niche overlap between mako shark, *Isurus oxyrinchus*, and Jumbo Squid, *Dosidicus gigas*, in the California Current." *California Cooperative Oceanic Fisheries Investigations Reports* 49 (2008).
Vince, Gaia. "How the world's oceans could be running out of fish." *BBC Future*, September 21, 2012.
Voss, Katrina. "Arctic Sea-Ice Loss Has Widespread Effects on Wildlife." *Penn State News*, August 1, 2013, http://news.psu.edu/story/283267/2013/08/01/research/arctic-sea-ice-loss-has-widespread-effects-wildlife.
Wade, Nicholas. "Adventures in Very Recent Evolution." *New York Times*, July 19, 2010.
———. "Genetic Data and Fossil Evidence Tell Differing Tales of Human Origins." *New York Times*, July 26, 2012.
Waggoner, Ben M., and Allen G. Collins. "The Cambrian Period." University of California Museum of Paleontology, November 22, 1994.
Wang, Lin-Fa. "Review of Bats and SARS." *Emerging Infectious Diseases* 12, no. 12 (December 2006).
Wang, Yifei. "A Cacophony in the Deep Blue: How Acidification May Be Deafening Whales." *Dartmouth Undergraduate Journal of Science*, February 22, 2009.
Ward, Peter. "What will become of *Homo sapiens*? Contrary to popular belief, humans continue to evolve . . ." *Scientific American*, January 2009.
Water in Anthropocene Conference, Bonn, Germany. "A majority on Earth face severe self-inflicted water woes within 2 generations." AAAS and *EurekAlert!* May 24, 2013.
Weiss, Kenneth R. "Beyond 7 Billion, Part 1: The Biggest Generation." *Los Angeles Times*, 2012.
———. "Beyond 7 Billion, Part 4: The China Effect." *Los Angeles Times*, 2012.
———. "Oxygen-Poor Ocean Zones Are Growing." *Los Angeles Times*, May 2, 2008.
"Were Dinosaurs Undergoing Long-Term Decline Before Mass Extinction?" American Museum of Natural History, October 26, 2012, http://www.amnh.org/our-research/science-news2/2012/were-dinosaurs-undergoing-long-term-decline-before-mass-extinction.
White, Tim D. "Human Evolution: The Evidence." In Brockman, John, ed. *Intelligent Thought: Science versus the Intelligent Design Movement*. New York: Vintage, 2006.
WHO Study Team. "Ebola haemorrhagic fever in Sudan, 1976." *Bulletin of the World Health Organization* 55, no. 2 (1978): 247–70. www.ncbi.nlm.nih.gov/pmc/articles/PMC2395561/.
WHO. "Human African trypanosomiasis: The history of sleeping sickness." www.who.int.
———. "Malaria, Fact sheet No. 94." *Media Centre*, http://www.who.int/mediacentre.
———. "Urgent action needed to prevent the spread of untreatable gonorrhoea." *Media Centre*, June 6, 2012, http://www.who.int/mediacentre.
Winerip, Michael. "The Second Act for Biosphere 2." *New York Times*, June 10, 2013.
Wing, Scott L., and Hans-Dieter Sues. "Mesozoic and Early Cenozoic Terrestrial Ecosystems." In *Terrestrial Ecosystems through Time*. Chicago: University of Chicago Press, 1992.
Wolf, Edward C. "Pictures from an Expedition: A Rapid Assessment of Rain Forests in Bolivia." *Orion Magazine*, Autumn 1991.
Wolverson, Roya. "Local Food Grows Up." *Time*, October 15, 2012.

Wynn, Thomas and Frederick L. Coolidge. "Into the mind of a Neanderthal." *New Scientist*, January 18, 2012.

"Xiao Xiao Receives Torch Lit by Dolly." *China Daily*, August 8, 2009, www.chinadaily.com.cn.

Young, Hilary S. "Declines in large wildlife increase landscape-level prevalence of rodent-borne disease in Africa." *Proceedings of the National Academy of Sciences* 111, no. 19 (May 13, 2014): 7036–41.

Index

About the Author

Michael Tennesen is a science writer who has written more than three hundred stories in such journals as *Discover*, *Scientific American*, *New Scientist*, *National Wildlife*, *Audubon*, *Science*, *Smithsonian*, and others. He was a Writer in Residence at the Cary Institute of Ecosystem Studies in Millbrook, New York, and a Media Fellow at the Nicholas School of the Environment and Earth Sciences, Duke University. He lives in the California desert near Joshua Tree National Park.